PERSON, SOCIETY AND VALUE
Towards a Personalist Concept of Health

Philosophy and Medicine

VOLUME 72

The titles published in this series are listed at the end of this volume.

PERSON, SOCIETY AND VALUE

Towards a Personalist Concept of Health

Edited by

PAULINA TABOADA
Assistant Professor, Pontificia Universidad Católica de Chile, Santiago, Chile

KATERYNA FEDORYKA CUDDEBACK
Arlington, Massachusetts, USA

PATRICIA DONOHUE-WHITE
Assistant Professor, Franciscan University of Steubenville, Steubenville, Ohio, USA

KLUWER ACADEMIC PUBLISHERS
DORDRECHT / BOSTON / LONDON

Library of Congress Cataloging-in-Publication Data is available.

ISBN 1-4020-0503-2

Published by Kluwer Academic Publishers,
P.O. Box 17, 3300 AA Dordrecht, The Netherlands

Sold and distributed in North, Central and South America
by Kluwer Academic Publishers,
101 Philip Drive, Norwell, MA 02061, U.S.A.

In all other countries, sold and distributed
by Kluwer Academic Publishers, Distribution Center,
P.O. Box 322, 3300 AH Dordrecht, The Netherlands

Printed on acid-free paper

Printed and bound in Great Britain by MPG Books Ltd., Bodmin, Cornwall.

TABLE OF CONTENTS

SECTION THREE / HEALTH AND SOCIETY

ACKNOWLEDGEMENTS

The authors of this volume acknowledge their gratitude to the Swiss National Science Foundation (SNF), which granted the interdisciplinary and international research project that lead to the collection of papers presented in this volume.

Our gratitude goes also to all those who collaborated with us in many different ways during this two-year project, and especially to Dr. H.T. Engelhardt, Jr., and his assistants Mark Cherry, Lisa Rasmussen, and Ana Smith Iltis, who, by their valuable comments and careful review work, made the publication of this volume possible.

We would also like to communicate to our readers that during the period of preparation of this work, one of its authors, Professor Armando Roa, died. We wish to express our sorrow over this great loss and extend to his family and collaborators our heart-felt condolences.

Paulina Taboada, M.D., M.phil.
Pontificia Universidad Católica de Chile
Santiago, Chile

ROCCO BUTTIGLIONE

PREFACE

From its foundation, the International Academy of Philosophy has considered a part of its task to be the inquiry into the practical relevance of philosophical truth, and participation in interdisciplinary investigations aimed at the solution of practical problems. This is perhaps in large part rooted in the phenomenological tradition that is such an important part of the philosophy that is taught at our Academy. Phenomenology is not only a specific philosophical methodology, it is also a way of thinking which can be characterized, among other things, by a general interest in formulating the problems of different disciplines in a clear and concrete way. This in turn is rooted in the general phenomenological conviction that the foundation of all reasoning, all deductions and all theories must be grounded in a foundation of clearly apprehended essential insights.

The attempt to come to grips with the problems of practical life can be divided into two different kinds of approaches. One starts with empirical data and tries to classify or organize them according to categories that are not grounded in the essence of things, but rather in a certain way of thinking. The other, not at all infrequently adopted, consists in predefining a framework – an abstract theory – and then trying to deduce from this abstract theory what should be done in a concrete case or the way in which the empirical data should be adjusted. If this "adjustment" proves to be difficult or even impossible, all the worse for reality, since this way of thinking leads the scientist to consider his first aim to be preserving the purity of his theoretical approach.

Phenomenology, instead, would like to give us a third way. It is convinced that in the empirical material itself there are elements of rationality, and recurring forms in the presentation of empirical data, which allow us to see what the fundamental essence of the phenomena is, and what the essential laws governing the phenomena of this kind are. This makes it possible, then, to investigate the essence of health in a way that allows us to find essential laws, essential insights that can guide the concrete action in this field. As a politician, I must add that such an approach can give us guidelines for concrete health care and welfare policies, which are badly needed today throughout the world, especially in the most advanced industrial countries.

It was with the background of such a conviction that we developed and submitted a research proposal to the Swiss National Science Foundation (SNF). The goal of our proposal was to take the WHO definition of health as our starting point, and through a cooperative effort of doctors and philosophers, work through the challenges which this (and indeed any) definition of health faces if it is to be both truly personalist and operational.

The collection of papers presented here is the product of a symposium that marked the mid-point of our two-year research. Written to cover topics we had found to be central in the course of the previous year, and re-written in light of the symposium discussions, the papers reflect not only the efforts of the interdisciplinary cooperation between medicine and philosophy, but also an international discussion of the problems central to the current debate about health. We are grateful to the Swiss National Science Foundation (SNF) for its support of our research, and to the symposium participants for their contribution to this research. And we are confident that this volume will indeed stand as a contribution to the important theoretical and practical problem of defining the nature and the value of health.

International Academy of Philosophy
Principality of Liechtenstein

KATERYNA FEDORYKA CUDDEBACK
AND PAULINA TABOADA

INTRODUCTION

I. ON THE TITLE OF THIS BOOK

In an attempt to introduce and situate the papers collected in this volume, it is worthwhile at the outset to explain a few things about the title of the volume: Why did we bring together the notions of person, society and value in a volume devoted to a critical analysis of the concept of health? What is meant to be conveyed by the term 'personalist', and by the term 'concept'? Was it necessary to include the open-ended expression 'towards' in our title?

The following remarks on the title should help in establishing a context for the discussions, and in providing a framework for uniting the various points of view contained in this volume.

A. The Use of the Terms 'Personalist' and 'Concept'

Taken *prima facie*, these terms seem not to require any discussion. This obviousness, however, is bought at the price of an uninformative vagueness: one could understand almost anything under the notion of a 'personalist' 'concept'. It will become evident from the papers collected here that to fix any precise meaning to the terms, and even more so to the terms in combination, is not an easy task.

The word 'personalist' is used very often in the contemporary discussion of health in opposition to biological, mechanistic, or functionalist approaches to human health. Models of health based on such approaches consider health primarily in terms of a physical organism or in terms of the functioning of this organism. Despite the various amounts of success such models have enjoyed, it has become clear that they fail ultimately to do justice to human health. This failure, however, would not be considered total by most, because to some extent these models have succeeded in capturing something important about the nature of health. The problem lies in limiting the characterization of health to what can be understood or explained in solely functionalist or biological terms, hence the tendency to consider these models reductionist, and not outright false.

P. Taboada, K. Fedoryka Cuddeback and P. Donohue-White (eds.), Person, Society and Value: Towards a Personalist Concept of Health, 1–15.
© 2002 *Kluwer Academic Publishers. Printed in Great Britain.*

They become reductionist because when applied to human health, they are forced to explain health in terms narrower than the full human reality.

What is the specifically human which these models do not capture? Most would agree that it is the moment of being a person, of being a subject who not only has a body, but also lives and consciously experiences this body and its states. Identifying the human being as a person carries with it the awareness that the human being is primarily a subject, and that this makes human health radically different from non-human: by being lived and consciously experienced, human life and health are no longer purely physical or biological realities.

This leaves us with a complexity in the notion of human health, born of the combination of a model of health as a biological, physical phenomenon, and a model that takes into account the dimension of subjectivity. It leaves us with the task of understanding the relation of subjectivity to physical and biological processes, and of the specific kind of being in which this union takes place. It challenges us also to recognize that identifying the specificity of human health requires the cooperation of many disciplines, as many as are necessary to do justice to this complex phenomenon.

Recognizing the need for a personalist model emerges, therefore, as only the first step in replacing inadequate theories of human health. The task requires the further step of making concrete what it means to be a person, and an embodied person as well. Then follows the step of identifying what belongs properly to health, and what falls outside of it. Each is in its own way a step in specifying the fundamental insight that a unique category, that of personhood, is necessary for an adequate characterization of human health. It is to be hoped that this process will yield in the end a 'concept' of human health. A few thoughts on how we use the term 'concept' will help us to explain further the goal of this volume.

As we have already noted, a given model of health is compatible with many different concepts of health. A concept differs from a model in that a concept can be considered as a concretization of a given model. It concretizes the model by putting forth a certain content, thereby delimiting some clear meaning and/or pointing to some concrete thing. A concept, then, is narrower than a model, more precise, and as a result, more difficult to formulate and easier to falsify or contradict. At the same time, it is more useful than a model, precisely because of the way it identifies or localizes.

On the other hand, a concept is richer or broader than a definition. As Lennart Nordenfelt notes in his *On the Nature of Health* (1987, p. 8), it is possible to have many definitions corresponding to, or attempting to capture, one concept. The reasons for such a multiplicity of definitions in relation to one concept could be many, such as the concept's ambiguity, or its complexity. This shows that a concept is more than a linguistic or propositional entity, while we could consider a definition to be such. A propositional formulation, while succeeding in perhaps pointing to an entity in such a way that it and only it answers to the definition, can in fact only point to its object, it cannot capture it. A concept, on the other hand, is able to 'capture' that which is meant, and could perhaps be characterized as an entity of the understanding, or as the mind's possession of the thing meant.

While definitions are certainly important and useful, their usefulness is directly proportional to the simplicity of the thing defined. The more complex the object, the less adequate is a definition in expressing what it is. Similarly, the more complex the object, the greater the need to go beyond a definition and understand the object in its own terms, or in other words, to develop a concept able to reflect and encompass this complexity. This is not to discourage the search for definitions, and in particular for a definition of health. It is only to point out that any definition presupposes understanding – having a concept – of that which is to be defined. Recognizing this dependence of definition on concept again shows the priority of clarifying and refining the concept. At the same time, it rescues definitions from their reductive potential by emphasizing that with every definition there is *more* than what is captured by the definition itself.

The notion of 'concept' itself is not an undisputed matter, as any basic course in logic soon makes clear. Of the debated issues, the one perhaps most crucial to the understanding of health is the question of to what exactly a concept refers. In a realist understanding, a concept is a concept of some *thing*, and through it we grasp the nature or essence of this reality. A nominalist understanding of concepts, on the other hand, sees concepts as reflective not of reality, but of an individual or collective structuring of the world or of the use of language. The resolution of this dispute depends ultimately on epistemological and ontological investigations, and the whole discussion could appear far removed from the concerns of medicine and the understanding of health.

Yet as the contemporary discussion as well as several of the papers included in this volume show, the ontological status of our conceptual correlate when talking about health is far from being settled. In speaking of health, does our concept refer to something real, or does it rather reflect an individual's or a society's interpretation of their physical or psychological states? The logical problem of the status of concepts and the ontological problem of the status of health on this point at least coincide, and a resolution of one question could help in resolving the other, and vice versa.

B. Which 'Personalism'?

It is evident that all the authors who contributed to this volume are committed to a personalist approach to understanding human health. On this point they stand in fundamental agreement. It soon becomes evident, however, that this fundamental agreement on the personal character of human health does not eliminate debate on the issue. An obvious point of discussion is the precise criteria for determining health itself. And yet, at least equally important is the question about what it means to be a person. The debate is no less a debate about the nature of the human person than it is a debate about the nature of health, and there can be as many definitions of human health as there are conceptions of what it means to be a human person.

Naturally, to grasp the uniqueness of the human person is not yet to grasp the specific nature of personal health. But it is certainly a first step, and it becomes evident that fundamental to the understanding of health is a certain anthropology, a certain theory of what it means to be human. Another way of putting this is to say that every theory of health presupposes (tacitly, if not explicitly) a theory of the person. Even a functionalist, mechanistic model, in prescinding from any consideration of persons, implies the possibility of such bracketing, and thus of a certain separability of person from body.

The theoretical implication of a specific anthropology, however, is not sufficient to extend the term 'personalist' to a given model. It seems more adequate, rather, to call those models 'personalist' which include the category of 'person' in the basic characterization of health. Proceeding as if one could understand health without reference to the personal character of the being whose health it is marks the fundamental difference between the various conceptions we have called reductionist above and beyond any personalist conception, which is based on the fundamental idea that one cannot understand human health without reference to human personhood.

In characterizing more precisely the position of this volume, it will be helpful to distinguish two major forms of personalism: realist and idealist. The former can best be understood in the context of traditional metaphysical realism, and the latter in terms of a modern version of idealism, for which personhood can be identified with the mind (or the presence of 'mental predicates'). Most of the contributions in this volume correspond to realist personalism, as represented by some contemporary Phenomenologists (e.g. Crosby, 1996; Scheler, 1973; Seifert, 1989, 1989a; Stein, 1994, 1994a) and the Polish Personalist School of Ethics (Wojtyla, Styczen & Szostek, 1979). According to this conception the person is the ontological ultimate and personhood the fundamental explanatory principle, hence the ultimacy attributed by most of the authors to personhood, both in value (the dignity of the person) and in being (the person as substance in the Aristotelian sense).

Moreover, personal existence and well-being are explored mainly through the phenomenological method, which attempts to capture in new ways the relation of the body to the person, a problem that has caused a long-standing ambiguity in personalist thought. The phenomenological investigations offered in this volume may therefore provide an impetus for new conceptions of personhood and human health. This approach has the advantage of advocating a systematic conception of the total person which combines surface experiences (subjective experiences of well-being) with deeper dimensions of the person (value and being).

It then becomes clear that this kind of realist personalism brings together the crucial concepts used in this collection, such as person, health, well-being, value, society, and so on. That a person is distinct from a mere thing and from other living beings, and that any human being, insofar as she is a person, is in consequence of this personal status to be treated in a special manner, has enormous implications not only for our understanding of what constitutes the health and well-being of the person, but also for our conception of what health care should comprise.

Thus, answering philosophical questions, such as those raised in this volume about health, is crucial for the solution of societal and political problems such as how to legislate and finance health care policies. Nevertheless, this task faces the challenges which any definition of health faces, if it is to be both truly personalist and at the same time operational.

C. An Ongoing Philosophical Endeavor

The purpose of this volume is to capture the essentials of human health and to propose the outlines for a personalist understanding of this concept, i.e., a conception that does justice to the personal nature of human beings by introducing dimensions that are essential to personal life and well-being, such as the realms of rationality, affectivity and freedom, the realms of meaning, values, morality, and spirituality, and the realms of social and interpersonal relations.

If the complexity of the matters considered is to be reflected in a concept, then the task of arriving at a concept of personal health emerges as no small task. And in light of this reflection, it becomes evident why the title is headed up with the word 'towards'. This 'towards' is expressive not only of the perhaps unavoidable inconclusiveness of the present volume, but also as a tribute to the richness and complexity of that with which it grapples.

II. ON THE CONTENT OF THIS BOOK

From the many areas that could be discussed, we have isolated and organized the individual papers into the consideration of three main themes: the nature of the human person, the nature of human well-being, and the implications of our previous analysis for the organization of society. The individual contributions need not have been grouped in this way, nor the sections classified in their present manner. Many in one section cover the focus of another, and various other concerns as well. At the same time, this presentation does reflect central emphases of the papers, as well as their dialogical, complementary character.

A. Section I. Health and the Human Person

The five papers included in this section direct our attention to the specifically personal dimensions of human health. All of them point to the fact that being a human person means being a subject who not only has a body, but also lives and consciously experiences this body and its states. In other words, identifying the human being as a person carries with it the awareness that the human being is primarily a subject, which makes human health radically different from non-human health: by being lived and consciously experienced, human life and health are no longer a purely physical or biological realities.

Reale's contribution argues that a holistic conception of the human person and her health was already present in ancient philosophy of medicine. Analyzing two of the Platonic dialogues (*Charmides* and *Timaeus*), the author shows that according to Plato, one cannot cure only a part of the body without curing the body as a whole, nor cure the evils of the body without also curing the evils of the soul. Drawing out the extreme consequences of Socrates' philosophy, Plato thought that it was not so much the body that one had to cure as the soul, for from the soul derives every form of good (and evil), in every sense and in every level, both in private and in public.

It is interesting to realize that in Plato's philosophy of medicine we encounter the anticipation of some Freudian concepts, such as the role of dreams in revealing our subconscious and the idea of the "Oedipus complex." By stressing this fact, Reale points out the extraordinary modernity of Plato's intuitions and the present significance of this doctrine.

Most contemporary efforts to provide a holistic conception of human health are based on the General Systems Theory. Taboada's paper deals with these efforts, asking whether or not they succeed in providing an intelligible framework for a personalist understanding of health. According to Taboada's inquiry, the conceptions of human health grounded on the General Systems Theory open an interesting path toward considering the human being not as a mere product or addition of parts, but as a whole which is much more than the sum of these parts, and therefore toward an understanding of the human being as a person. With regards to the concept of health, these theoretical models offer a way in which the health of a person can be disturbed by alterations occurring in the biochemical order, in personal interactions, in the family atmosphere, or in the macrosocial order. Accordingly, these models aim at accounting for the fact that a dysfunction at any of these levels can affect the health of the person as a whole. Nevertheless, Taboada concludes that despite its interesting contributions toward a more comprehensive notion of human health, these new conceptions entail important anthropological and epistemological problems that need to be overcome if we want to secure a cogent, personalist conception of health and health care.

This challenge is assumed by Ide, whose contribution intends to offer a 'third way' between the two classical models of health (reductionist and holistic). After undertaking an in-depth critical analysis of the anthropological assumptions upon which the two classical conceptions are grounded, Ide proposes an original approach to the concept of health based on an anthropology and ontology of the 'gift'. His main idea is that the act of giving, understood as a dynamic process with three stages - the reception of the gift (the gift for the self); the appropriation of the gift (the gift to the self); and the offering of the gift (the gift of the self) – is central to an authentically personalist understanding of human health.

Ide shows the way in which his new approach to health is able to shed light on some of the positive and deficient aspects of the two traditional conceptions of health. Indeed, he asserts that his definition of health as appropriation of the body cogently accounts for the true insights of the systems conception of the unified body, doing justice to the key role attributed to the spontaneous dynamism of the body in the systems approach to health. He also intends to situate in their proper place the contributions of the biomedical models, stressing the role of technological interventions in the care for health. The link between two conceptions of health that have been traditionally opposed to each other is accounted for by this author through a reference to Aquinas' hierarchical articulation of the 'main' and the 'adjuvant' causes. Applying this distinction to the process of recovering health, Ide suggests that the intrinsic dynamism of the human body are the main cause of recovery, while the medical art plays an adjuvant role.

Ide's application of the 'logic of the gift' to understanding the health of the human person leads to a strong criticism of the contemporary 'idolization' of health. His conclusion is that "if we want to understand what is ultimately at stake with health we must appeal to a properly religious vision or rather to a reinsertion of health within a complete vision of reality that would integrate the three regional ontologies: God, man and nature" (p. 72, in this volume).

In his article, "The Concept of Mental Health," Roa proposes a similar approach. He first criticizes current understandings of mental health grounded on biological, psychosocial and antipsychiatric models. Based on his vast clinical experience, this psychiatrist undertakes an anthropological-phenomenological analysis of the characteristics which indicate mental disorders, and proposes an original concept of mental health as "the integrity of the mind, the constant possibility of fulfilling what a human life in accordance with its inner nature demands for its realization in every stage of life: infancy, adolescence, youth, maturity and old age" (p. 96, in this volume). Central to Roa's conception of mental health is the idea of human freedom; to be mentally sane means to have the "possibility of realizing the desire to give the highest possible fulfillment to what is proper to human nature – one's own nature as much as the nature of those around us – making it our concern that we give the best of ourselves to others and that we receive the best of others, in an active interpersonal relationship" (p. 97).

In fundamental agreement with Roa, and applying the phenomenological method in a systematic way, Seifert explores in depth the relations between the body and the person. He establishes a threefold relationship of human health to pre-biological, biological and mental dimensions of the human person, suggesting that these different dimensions are all part of specifically human health only in virtue of their insertion and integration into a cogent anthropology. In spite of including many aspects of conscious life and aesthetic values in a broader concept of human health, Seifert does not remain captured by subjectivism, but insists on the objective nature of a personalist concept of health grounded in the intelligible nature of the human being that includes both empirically discovered and philosophically grasped aspects of the human person.

B. Section II. Health and Human Well-being

The second section of the volume comprises three articles concerning mainly the relations between health, value, and human well-being. Though they differ from each other in significant ways, the three articles taken together illustrate the complexity of the relations between health, goodness, and well-being and even in their disagreement, contribute to the clarification of the phenomena in question.

The first contribution, by H. Tristram Engelhardt Jr., assumes epistemological (but not metaphysical) scepticism and on this basis argues for a plurality of well-beings in a post-modern world. The second paper, by Manuel Lavados, and the third, co-authored by Patricia Donohue-White and Kateryna Fedoryka Cuddeback, assume epistemological (and metaphysical) realism and, in differing ways, argue for a general notion of well-being rooted in human nature.

In his article, "Health, Disease and Persons: Well-Being in a Post-Modern World," Engelhardt argues that there is no generic well-being but rather numerous competing accounts of the good life and human flourishing. Engelhardt labels the situation post-modern and argues on the basis of epistemological scepticism that one cannot choose in a principled fashion on the basis of a general secular sound rational argument among these competing accounts. This plurality of accounts of well-being discloses the possibility of a plurality of social understandings of the goals of medicine and invites the emergence of alternate, parallel health care delivery systems.

The weight that Engelhardt attributes to community in the constitution of meaning and knowledge cannot be overrated. Outside of a particular community with a particular moral vision, Engelhardt argues, individual persons are naked: "they have no meaning they themselves can establish as binding" (p. 59, in this volume). And outside of a particular community, it is impossible to realize the goods of health care since such goods are constituted as goods only within a particular community.

Outside of the relations which constitute specific communities, Engelhardt concludes, people can be united by consent and agreement: "consent to affirm particular understandings of well-being; agreement to collaborate within particular communities directed to particular understandings of well-being" (p. 160, in this volume). In this general realm, disease, health and well-being are largely social constructs.

Lavados' contribution on the "Empirical and Philosophical Aspects of a Definition of Health and Disease" presupposes a realist foundation for his analysis and gives far greater weight to generic notions of health and disease. Lavados acknowledges both objective and subjective dimensions of health and well-being and examines the interrelation of these two dimensions through an analysis of the biomedical and holistic models of health. He argues that rather than being opposed to one another, each model emphasizes a different dimension of the human being and therefore offers a different understanding of health. The biomedical model approaches the human being as a member of a specific animal species, emphasizes disease, and establishes the norm of health in terms of whether or not the individual functions according to the expectations for the biological species. In contrast, the holistic model approaches the human being as a person, thereby emphasizing subjective experience, social dimensions, and human freedom, and tends to define health in these terms.

Lavados argues that by introducing the philosophical concept of *nature*, a 'conceptual bridge' can be forged between these two models and a definition of health can be achieved which takes the insights of both into account. To have a *nature* is to be oriented and to function according to some end. Being healthy corresponds to functioning in conformity with the natural design of the organism; disease consists in a state wherein the organism is unable to exercise one or more of the functions typically performed by a member of the species.

Having established definitions of health and disease in terms of the concept of nature, Lavados returns to a consideration of the holistic and biomedical models. According to him, holistic models emphasize the subjective dimensions of disease and define health in terms of happiness. In contradistinction, biomedical models emphasize structure and function. In spite of this difference, holistic and biomedical models are not so far from one another, argues Lavados. Holistic models imply that the human being is a natural being since the individual is intrinsically directed to a state (happiness) common to all human beings and there are objective conditions which are necessary to achieve this state. Though Lavados does not say so explicitly, he seems to imply that while the biomedical model affords a more adequate definition of health and disease, the holistic model recognizes and emphasizes dimensions of the human person which go beyond health (e.g., the whole realm of human freedom) and thus points to a fuller understanding of the human person.

The third contribution, "The Good of Health and its Normativity for Medicine," by Patricia Donohue-White and Kateryna Fedoryka Cuddeback, critically examines the general subjectivist theory of the good and well-being, referred to as the desire-satisfaction theory, which underpins dominant contemporary theories of health. Donohue-White and Cuddeback concede that a comprehensive understanding of human health includes both value and normative dimensions. They sympathize with what they identify as a primary motive for adopting theoretical subjectivism: the concern with placing the individual person's well-being at the center of medical practice rather than general physiological functions and species goals (which they call the 'personalist motive'). Nevertheless, they reject desire-satisfaction theory as theoretically incoherent, impracticable as a foundation for the norms of praxis, and finally, inadequate to the experience of good and well-being.

The authors offer an objectivist understanding of the good and well-being which, they argue, does not suffer from incoherence and provides a foundation for the norms of praxis. They also assert that such an objectivist account is more adequate to experience and grounds a richer notion of the human person. Consequently, it provides a more adequate foundation for a personalist approach to health care.

Donohue-White and Cuddeback proceed from a realist foundation. They would grant Engelhardt a certain plurality of well-beings once one moves to the level of the specific, content-full well-being of individual persons. Nonetheless, the authors assert that there is a generic well-being rooted in human nature which forms the 'background' of all specific well-beings, and far from being an 'empty function', the concept of generic well-being serves as the foundation for realist conceptions of human health. With reference to phenomenological analysis, the authors argue that the content of well-being is causally independent of desires and claim that the well-being of the human person is fundamentally determined by the nature of the person. They suggest that a general concept of health can be derived from this concept of generic well-being.

Nevertheless, in the case of the human person, they argue, one must consider more than a general nature that organically unfolds. One must also consider dimensions that are essential to personal life and well-being such as the realms of rationality, affectivity, and freedom, the realms of meaning, values, morality, and spirituality, and the realms of social and interpersonal relations. The authors acknowledge the complexity and manifoldness of the human person and in consequence eschew simple definitions of the person and personal well-being. They conclude by arguing that an objectivist account of human health and well-being is not only theoretically coherent, but also allows for the transition from theory to praxis, which is the main theme of the final section.

C. Section III. Health and Society

The two contributions of the last section bring out the significance of resolving the philosophical questions discussed in the first two sections for the organization of health care systems and for public policy. Starting from a critical analysis of the WHO definition of health as "a state of complete physical, mental and social well-being, and not merely the absence of disease and infirmity" (WHO, 1946), both papers point out that this definition requires further specification if it is to be practically useful.

Van Spijk criticizes both the content and the methodology of the WHO definition of health. A review of the literature on this definition shows that many criticisms have been offered. Most of them deal mainly with its content. Van Spijk's original contribution consists in pointing out the methodological deficiencies of the WHO definition. While showing some of the problems with the manner in which this definition was drafted by the WHO preparatory committee, this author derives interesting conclusions about the adequate way of undertaking such a delicate task. He argues that answering the question about the nature of health demands, besides a "careful and respectful study of what others in the past have thought and written on the subject", an "intensive, long-term research done by scientists who are deeply motivated to solve this problem, but who are also patient enough to let the question be solved within our time and our social context" (p. 217, in this volume).

In an interesting attempt to systematize the present knowledge on the concept of human health, van Spijk introduces a four-axis model based on General Systems Theory. These four axes include:

1. Three dimensions of the object: state, function and goals.
2. Five levels of the system: basic elements, cell, organ, organism, environment.
3. Three levels of being: physical, biological, self-aware.
4. Two perspectives of describing: objective, subjective.

Although this systematization intends to shed some light on the question about the nature of human health, the author consciously does not attempt to formulate any definition. He rather chooses to sketch out the difficult and complex paths toward a good definition. His purpose is to provide some concepts and methodologies that may be helpful in the search for a coherent and practically applicable definition of health.

Buttiglione and Pasquini approach the notion of well-being contained in the WHO definition of health from the point of view of the politician. They explore the significance of resolving the philosophical question about how well-being is to be understood for public policy. One of the greatest challenges governments face with regard to health is that of financing health care. While this is a political and economic problem, the authors claim, three philosophical issues underlie this problem. In presenting the problem, Buttiglione and Pasquini invite the reader to perform a 'paradigm shift'. They ask us to consider the importance of the problem not from the perspective of the physician or the patient, but from that of the taxpayer, i.e., the one who ultimately finances health care.

The first question to be considered in attempting to solve the problem of financing health care is what constitutes 'health'. The fundamental problem they see with the WHO definition is that it allows for various interpretations. It is necessary to specify what is meant by the term 'well-being'. Three types of approaches to this specification are pointed out: emotionalist, materialist, and realist. The realist approach, according to the authors, besides being more adequate from a philosophical standpoint, is also the one best-suited to the pragmatic demands arising from the point of view of the tax-payer. Of the three approaches, it best provides a sharply delineated definition of health, in which neither the subjective nor the objective components of health are emphasized at the cost of the other, and in which there is a safeguard against the arbitrariness of an emotionalist understanding of well-being, which would give rise to endless demands, because of the limitless expectations of the subjectively felt needs.

Secondly and thirdly, it must be asked whether there is a right to health, and, if so, what the corresponding duties are, and to whom they are addressed. Answering these questions goes beyond understanding what health itself is. It incorporates the question about health, but goes beyond it because it requires identifying whether this right exists, whether it is a positive or a negative one, and finally whether it is addressed to the individual, civil society, the government, or to all three in different ways. A realistic answer to these questions, according to Buttiglione and Pasquini, is one which recognizes that all three levels - individual, society and government - must cooperate in the securing of health. Any attempt to lay the burden on any one of these in isolation does not correspond to "the reality of today's world," "the essence of social responsibility" or "the essence of health" (p. 237, in this volume).

Buttiglione and Pasquini conclude by making the point that "in order to make good laws, we must make clear what is the finality, the goal, of the laws we make. And, if the goal is health, we must know what health is" (p. 238, in this volume). The best approach to finding a politically realistic solution, they maintain, is to begin with a philosophical realism.

Summing up the content of this book one may say that each of the contributions here collected open a door to considering the wide, fascinating and challenging philosophical landscape which starts to unfold once the question about the nature of health is asked. The general conclusion of the volume is that answering philosophical questions, such as those raised here about the essence and the value of human health, is crucial for the solution of political problems such as how to legislate health care policy. We hope that reading this book will provide a stimulus for those who would like to engage in taking a further step on the difficult and complex path toward a personalist concept of health.

International Academy of Philosophy
Principality of Liechtenstein

REFERENCES

Crosby, J.: 1996, *The Selfhood of the Human Person*. The Catholic University of America Press, Washinghton D.C.

Nordenfeldt, L.: 1987, *On the Nature of Health*, Reidel, Dordrecht.

Scheler, M.: 1973, *Formalism in Ethics and Non-Formal Ethics of Value. A New Attempt Toward the Foundation of an Ethical Personalism*. M. Frings & R. Funk (trans.), Northwestern University Press, Evanston.

Seifert, J.:
—— 1989: *Das Leib-Seele-Problem und die gegenwärtige philosophische Diskussion. Eine systematisch-kritische Analyse*. Wissenschaftliche Buchgesellschaft, Darmstadt.
—— 1989a: *Essere e Persona. Verso una Fondatione Fenomenologica di una Metafisica Classica e Personalistica*. Vita e Pensiero, Milano.

Stein, E.:
—— 1994: *Was ist der Mensch?* Edith Steins Werke, Bd. XVII, Herder, Feiburg i.B.
—— 1994a: *Der Aufbau der menschlichen Person*. Edith Steins Werke, Bd. XVI, Herder, Feiburg i.B.

Wojtyla, K., Szostek, A. & Styczen, T.: 1979, *Der Streit um den Menschen. Personaler Anspruch des Sittlichen*. Butzon & Bercker, Kevelaer.

World Health Organization (WHO): 1946, *Proceedings and Final Acts of the International Health Conference, held in New York from 19 June to 22 July 1946. Off. Rec. Wld Hlth Org.*, 2, 1-143.

SECTION ONE

HEALTH AND THE HUMAN PERSON

GIOVANNI REALE

ACCORDING TO PLATO, THE EVILS OF THE BODY CANNOT BE CURED WITHOUT ALSO CURING THE EVILS OF THE SOUL

Plato knew well the medicine of his time. He also had his own form of a philosophy of medicine containing intuitions of extraordinary significance and a core of truth which in certain respects remains largely valid today. I will investigate this point in two of Plato's dialogues: the *Charmides*, which dates from his youth, and the *Timaeus*, his last published work.

I. ONE CANNOT CURE WITHOUT CURING THE WHOLE MAN

A. *One Cannot Cure Only a Part of the Body Without Curing the Body as a Whole*

In the *Charmides* medicine is spoken of in terms of a rather well-defined metaphor which represents in concise form the principal line of Plato's argument. It is therefore worthwhile to understand the metaphor properly.

The young Charmides has a bad headache and Socrates presents himself to Charmides as possessing a remedy with which to cure him:

[The remedy] was a certain leaf, but there was a charm to go with the remedy; and if one uttered the charm at the moment of its application, the remedy made one perfectly well; but without the charm there was no efficacy in the leaf (*Charmides*, 155E).

The significance of the *leaf* presents no difficulty inasmuch as it symbolizes the medicine or the drug, as the text says expressly. In contrast, the meaning of the image of the *charm* is much more complex. Let us follow the reasoning Plato uses to make the image understandable to the reader.

In the first place, one has to understand the structural link which joins each part of the human body with the body itself in its wholeness. This structural link is essential to such a degree that the mode of being of the part depends strictly on the mode of being of the whole. Consequently,

P. Taboada, K. Fedoryka Cuddeback and P. Donohue-White (eds.), *Person, Society and Value: Towards a Personalist Concept of Health*, 19–31.
© 2002 *Kluwer Academic Publishers. Printed in Great Britain.*

the part cannot be treated in itself and for itself, independently of the whole. Thus, one will not be able to treat the head without curing the whole head, and one will not be able to cure the head without curing the body as a whole. This, says Plato, is what the most authentic physicians do (and we would add, what the physicians of today who too easily lose themselves in extreme specialization and subspecialization should also do). Let us read the beautiful text:

> I dare say you have yourself sometimes heard good doctors say, you know, when a patient comes to them with a pain in his eyes, that it is not possible for them to attempt a cure of his eyes alone, but that it is necessary to treat his head too at the same time, if he is to have his eyes in good order; and so again, that to expect ever to treat the head by itself, apart from the body as a whole, is utter folly (*Charmides*,156B-C).

What then must the good physician do? He must seek to cure the part, but always with his gaze turned to the totality of the organism.

B. The Evils of the Body Cannot Be Cured Without Also Curing the Soul

But notice now the surprising twist which Plato presents us with after having explained this beautiful concept (which is, by the way, a canon of good Greek medicine). He tells the story of how Socrates, finding himself in a military camp, met a Thracian physician, a student of Zalmoxis (who was venerated like a divinity, and to whose disciples the therapeutic ability of procuring immortality was attributed), who revealed to Socrates the true art of curing the body.

The Thracian physician is obviously a dramaturgical mask through which Plato communicates his own thought in an emblematic way. His argument is the following. The Greek physicians were right in maintaining that one cannot cure a part of the body without curing the body as a whole. But the body is not man in his integral wholeness but rather only one part of him; the totality of man is his body and his soul. Thus, *just as one cannot cure a part of the body without curing the whole of the body, analogously one cannot cure the body without also curing the soul.* But the Greek physicians ignored this and consequently it escaped their attention that most diseases can be cured only if one not only cures the whole body, but also the whole man, that is, the soul as well as the body.

It is precisely from the soul that the greatest evils come to man, as well as the greatest goods:

> But Zalmoxis, our king, who is a god, says that as you ought not to attempt to cure eyes without head, or head without body, so you should not treat body without soul; and this was the reason why most maladies evaded the physicians of Greece - that they neglected the whole, on which they ought to spend their pains, for if this were out of order it was impossible for the part to be in order. For all that was good and evil, he said, in the body and in man altogether was sprung from the soul, and flowed along from thence as it did from the head into the eyes. Wherefore that part was to be treated first and foremost, if all was to be well with the head and the rest of the body. And the treatment of the soul, my wonderful friend, is by means of certain charms (*Charmides*, 156E- 1577A).

Plato then makes an explicit statement concerning the "charms." They refer to philosophy which educates man in "temperance," that is, in that virtue which teaches him what the *right measure* consists in and how to practice it:

> These charms are words of the right sort: by the use of such words is temperance engendered in our souls, and as soon as it is engendered and present we may easily secure health to the head, and to the rest of the body also (*Charmides*, 167A-B).

II. THE ORIGIN OF HUMAN DISEASES ACCORDING TO THE *TIMAEUS*

The concept of the right measure implicit in the concept of the virtue of temperance is developed by Plato in a final section of the *Timaeus* dedicated to the diseases of the body and the soul and their cure. Here it is extended to all levels of man and strictly connected to health.

A. The Three Groups of Bodily Diseases and Their Origin According to the Timaeus

Diseases are divided into three fundamental groups. The first group originates in various forms of disturbances which occur in the very composition of the four elements which constitute the body (water, air, earth and fire). The excess or defect in the composition of these elements, or a shifting of one of them *against nature*, disrupts the right measure, and thus gives rise to disturbances which generate diseases. Notice how Plato affirms the fundamental concept of his reasoning concerning the right measure:

> For, as we maintain, it is only the addition or subtraction of the same substance from the same substance in the same order and in the same manner and in due proportion which will allow the latter to remain safe and sound in its sameness with itself. But whatsoever oversteps any of these conditions in its going out or its coming in will produce alterations of every variety and countless diseases and corruptions (*Timaeus*, 82B).

The second group of diseases derives from disturbances which take place in the secondary compositions, i.e., from disturbances of the compositions from which derive the marrow, the flesh, the bones, the nerves, etc. Diseases arise when the compositions which constitute these parts of the body and their functions occur in a mode which is not in conformity with the laws of nature but opposed to them. Plato describes the most serious source of these diseases in the following way:

> And the most extreme case of all occurs whenever the substance of the marrow becomes diseased either from deficiency or from excess; for this results in the gravest of diseases and the most potent in causing death, inasmuch as the whole substance of the body, by the force of necessity, streams in the reverse direction (*Timaeus*, 84C).

The third group of diseases is connected with the air, the phlegm and the bile. Specifically, they are connected to the quantity of air which enters into the body in greater or lesser quantity with respect to that which is due, and to the tensions which follow from this. And so, also in the case of the phlegm and the bile, the diseases derive from their greater or lesser quantity with respect to that which is due, from their *excess* or *defect*, that is, from the disruption of the right measure.

B. Diseases of the Soul and their Origin

Diseases of the soul derive from a want of intelligence, either madness or ignorance. Many such diseases do not derive only from the soul itself, but also from those physical conditions of the body which influence it.

The first group of evils derives from an excess of either pleasure or pain, inasmuch as this excess deprives a man of the proper way of seeing things and of acting accordingly.

> For when a man is overjoyed or contrariwise suffering excessively from pain, being in haste to seize on the one and avoid the other beyond measure, he is unable either to see or to hear anything correctly, and he is at such a time distraught and wholly incapable of exercising reason (*Timaeus*, 86B-C).

The second group of evils comes from an excess of seed which forms itself in the marrow, producing excessive pleasures and pains, and passions without limit. The man struck by this excess, finds himself in the following condition: "He ... comes to be in a state of madness for the most part of his life because of those greatest of pleasures and pains" (*Timaeus*, 86C-D).

At this point Plato expresses an exquisitely Greek moral judgment of extraordinary Hellenic magnanimity and of a strongly Socratic flavor. Of such a type of man he says the following:

> He ... keeps his soul diseased and senseless by reason of the action of his body. Yet such a man is reputed to be voluntarily wicked and not diseased; although, in truth, this sexual incontinence, which is due for the most part to the abundance and fluidity of one substance because of the porosity of the bones, constitutes a disease of the soul. And indeed almost all those affections which are called by way of reproach "incontinence in pleasure," as though the wicked acted voluntarily, are wrongly so reproached; for no one is voluntarily wicked, but the wicked man becomes wicked by reason of some evil condition of body and unskilled nurture, and these are experiences which are hateful to everyone and involuntary (*Timaeus*, 86 D-E).

Plato also holds that certain diseases derive from a bad education. In the first place, certain diseases derive from bad city administration where bad discourses are spread publicly and privately, and, consequently, spaces for the diffusion of sound doctrines which could heal those evils are lacking:

> Furthermore, when, with men in such an evil condition, the political administration also is evil, and the speech in the cities, both public and private, is evil; and when, moreover, no lessons that would cure these evils are anywhere learnt from childhood, - thus it comes to pass that all of us who are wicked become wicked owing to two quite involuntary causes (*Timaeus*, 87B).

The second cause of certain diseases is to be found in the bad education given to the young. Consequently, those truly responsible for these diseases are parents and educators.

> And for these we must always blame the begetters more than the begotten, and the nurses more than the nurslings; yet each man must endeavor, as best he can, by means of nurture and by his pursuits and studies to flee the evil and to pursue the good (*Timaeus*, 87B).

III. THE BASIC RULES FOR THE CURE OF THE BODY AND SOUL

What then is the general criterion which Plato indicates for the cure of the body and the soul? The basic criterion is the same criterion which, as it were, constitutes the general foundation of Platonic thought: it is precisely that which renders all things good and beautiful, that is, the right measure.

A. *Role of the Right Measure*

One should keep in mind that, for Plato, the *supreme good* coincides with the *supreme measure* of all things, and that at all levels, each form of good is born from the precise function which the One-Measure exerts on the disordered multiplicity, bringing the disorder to order in various ways.[1] We can see how this criterion is summarized in the section of the *Timaeus* on medicine:

> All that is good is fair, and the fair is not void of due measure; where also the living creature that is to be fair must be *symmetrical*. Of symmetries we distinguish and reason about such as are small, but of

the most important and the greatest we have no rational comprehension (*Timaeus*, 87C-D).

And, in the case of man, the most important right measure, the one on which health depends, is that which must be established between body and soul, while the lack of measure between these two components of man leads to great sicknesses:

> For with respect to health and disease, virtue and vice, there is no symmetry or want of symmetry greater than that which exists between the soul itself and the body itself. But as regards these, we wholly fail to perceive or reflect that, whenever a weaker and inferior type of body is the vehicle of a soul that is strong and in all ways great, - or conversely, when each of these two is of the opposite kind, - then the creature as a whole is not fair, seeing that it is unsymmetrical in respect of the greatest symmetries; whereas a creature in the opposite condition is of all sights, for him who has eyes to see, the fairest and most admirable (*Timaeus*, 87D).

When a soul is clearly superior to the body, excitable and fiery, it fills the body with various diseases which physicians often are unable to recognize and attribute to external causes. In contrast, when a body is great and robust and the soul which is in it is small and weak, the bodily desires have pre-eminence and cloud the desires of the soul. Consequently, the soul becomes obtuse and ignorant. Here is Plato's therapeutic proposal:

> From both these evils the one means of salvation is this - neither to exercise the soul without the body nor the body without the soul, so that they may be evenly matched and sound of health. Thus the student of mathematics, or of any other subject, who works very hard with his intellect must also provide his body with exercise by practicing gymnastics, while he who is diligent in molding his body must, in turn, provide his soul with motion by cultivating music and philosophy in general, if either is to deserve to be called truly both fair and good (*Timaeus*, 88B-C).

The concept expressed in the famous Latin maxim *mens sans in corpore sano* ("a healthy soul is in a healthy body") was perfectly formulated by Plato himself.

The body is cured, in particular, with gymnastics and by engaging in regular movements with specific ends. From a certain point of view, Plato

seems have had more esteem for gymnastics as a cure for the body than
he had for medicine, inasmuch as gymnastics prevents disease, while
medicine intervenes only after diseases have already broken out. Man
must trouble himself much more with the prevention of diseases than with
their cure, for diseases arise above all when one neglects their prevention.

B. On the Use of Medication

In the *Timaeus* Plato also clarifies his thoughts on the use of drugs. In his
judgment, one should use drugs with caution and with great moderation
since, in certain cases, the harm is greater than the benefit. Note here
Plato's own words, which still possess a certain force against the
prevailing abuse of drugs by people today:

> For no diseases which do not involve great danger ought to be irritated
> by drugging. For in its structure every disease resembles in some sort
> the nature of the living creature. For, in truth, the constitution of these
> creatures has prescribed periods of life for the species as a whole, and
> each individual creature likewise has a naturally predestined term of
> life [...]. With respect to the structure of diseases also the same rule
> holds good: whenever anyone does violence thereto by drugging, in
> despite of the predestined period of time, diseases many and grave, in
> place of few and slight, are wont to occur (*Timaeus*, 89B-C).

C. The Care of the Soul Must Have Priority Over Everything Else

According to Plato, before and more importantly than the care of the
body, man must devote himself to the care of the soul, which must guide
the body "so that it may be as fair and good as possible for the work of
guidance" (*Timaeus*, 89D).

Above all, man must take care of the most important part of the soul,
that is, the rational soul, because God has put it in us as our *daemon,* and
it is the link which reunites us with the divine: to cultivate this soul, the
daemon which is in us, and to keep it ordered, is more than any other
thing what renders human life happy. Let us read the truly emblematic
passage which contains, as it were, the culmination of Plato's address:

> And as regards the most lordly kind of our soul, we must conceive of it
> in this wise: we declare that God has given to each of us, as his
> daemon, that kind of soul which is housed in the top of our body and

which raises us - seeing that we are not an earthly but a heavenly plant - up from earth towards our kindred in the heaven. And therein we speak most truly; for it is by suspending our head and root from that region whence the substance of our soul first came that the Divine Power keeps upright our whole body.

Whoso, then, indulges in lusts or in contentions and devotes himself overmuch thereto must of necessity be filled with opinions that are wholly mortal, and altogether, so far as it is possible to become mortal, fall not short of this in even a small degree, inasmuch as he has made great his mortal part. But he who has seriously devoted himself to learning and to true thoughts, and has exercised these qualities above all his others, must necessarily and inevitably think thoughts that are immortal and divine, if so be that he lays hold on truth, and in so far as it is possible for human nature to partake of immortality, he must fall short thereof in no degree; and inasmuch as he is for ever tending his divine part and duly magnifying that daemon who dwells along with him, he must be supremely blessed (*Timaeus*, 90A-C).

He hopes in vain who holds that he can achieve the true health of man by curing only or primarily the body since many of the evils and many of the goods of man derive from the soul. Thus, if you want to cure yourself, begin precisely with curing your soul.

IV. THE ANTICIPATION OF FREUDIAN CONCEPTS IN PLATO

Whoever has followed me to this point will have certainly become aware of the extraordinary modernity of some of the thoughts expressed by Plato. In conclusion, I would like to indicate some of these surprising foreshadowings.

André Gide asserted that he drew no more from Freud than what he had already found in Plato. In effect, if one prescinds from the provocative tone of Gide, it is true that Plato emphasized some ideas which psychoanalysis would first bring to prominence in the works of Freud. Two ideas in particular are here put in relief. They are contained in the following passage of the *Republic:*

– In the matter of our desires I do not think we sufficiently distinguished their nature and number. And so long as this is lacking our inquiry will lack clearness.

– Well, will our consideration of them not still be opportune?

– By all means. And observe what it is about them that I wish to consider. It is this. Of our unnecessary pleasures and appetites there are some lawless ones, I think, *which probably are to be found in us all, but which, when controlled by the laws and the better desires in alliance with reason, can in some men be altogether got rid of, or so nearly so that only a few weak ones remain, while in others the remnant is stronger and more numerous.*

– What desires do you mean?

– *Those that are awakened in sleep when the rest of the soul, the rational, gentle and dominant part, slumbers, but the beastly and savage part, replete with food and wine, gambols and, repelling sleep, endeavors to sally forth and satisfy its own instincts.* You are aware that in such case there is nothing it will not venture to undertake as being released from all sense of shame and all reason. *It does not shrink from attempting to lie with a mother in fancy or with anyone else, man, god, or brute.* It is ready for any foul deed of blood; it abstains from no food, and, in a word, falls short of no extreme of folly and shamelessness.

– Most true.

– But when, I suppose, a man's condition is healthy and sober, and he goes to sleep after arousing his rational part and entertaining it with fair words and thoughts, and attaining to clear self-consciousness, while he has neither starved nor indulged to repletion his appetitive part, so that it may be lulled to sleep and not disturb the better part by its pleasure or pain, but may suffer that in isolated purity to examine and reach out toward and apprehend some of the things unknown to it, past, present, or future, and when he has in like manner tamed his passionate part, and does not after a quarrel fall asleep with anger still awake within him, but if he has thus quieted the two elements in his soul and quickened the third, in which reason resides, and so goes to his rest, you are aware that in such cases he is most likely to apprehend truth, and the visions of his dreams are least likely to be lawless.

– I certainly think so.

– This description has carried us too far, but the point that we have to notice is this, *that in fact there exists in every one of us, even in some*

reputed most respectable, a terrible, fierce, and lawless brood of
desires, which it seems are revealed in our sleep (Republic IX, 671A-
572B).

I believe that in reading this passage attentively everyone will recognize
two key points of psychoanalysis: first, the role of dreams in revealing
our subconscious, and secondly, the "Oedipus complex," that is, the
representation of the desire for an incestuous union with the mother.

V. CONCLUDING REMARKS

Naturally at this point, I could direct the discourse to the way in which
Plato sought to cure the soul of man with the search for truth, but the
theme which I had to treat here is itself exhausted.

In these thoughts of Plato which we have examined, Plato draws out
the extreme consequences of Socrates' philosophy. For if there is one
message which Socrates wanted to teach with his philosophy it is this:
that it is not so much the body that one has to cure as the soul, for from
the soul derives every form of good (and evil), in every sense and at all
levels, both in private and in public.

Let us read the passage which summarizes the thought of Socrates,
taken from his defense:

Gentlemen, I am your very grateful and devoted servant, but I owe a
greater obedience to God than to you, and so long as I draw breath and
have my faculties, I shall never stop practicing philosophy and
exhorting you and elucidating the truth for everyone that I meet. I shall
go on saying, in my usual way, My very good friend, you are an
Athenian and belong to a city which is the greatest and most famous in
the world for its wisdom and strength. Are you not ashamed that you
give your attention to acquiring as much money as possible, and
similarly with reputation and honor, and give no attention or thought to
truth and understanding and the perfection of your soul?

And if any of you disputes this and professes to care about these
things, I shall not at once let him go or leave him. No, I shall question
him and examine him and test him; and if it appears that in spite of his
profession he has made no real progress toward goodness, I shall
reprove him for neglecting what is of supreme importance, and giving
his attention to trivialities. I shall do this to everyone that I meet,

young or old, foreigner or fellow citizen, but especially to you, my fellow citizens, inasmuch as you are closer to me in kinship. This, I do assure you, is what my God commands, and it is my belief that no greater good has ever befallen you in this city than my service to my God. For I spend all my time going about trying to persuade you, young and old, to make your first and chief concern not for your bodies nor for your possessions, but for the highest welfare of your souls, proclaiming as I go, Wealth does not bring goodness, but goodness brings wealth and every other blessing, both to the individual and to the state (*Apology*, 29D-30B).

I do not believe that there could be a more significant passage than this: the care of the soul is that from which alone derives the true health of man.

Plato is the philosopher who has taken this "cure of the soul" to the highest level in the ancient world. The ancients themselves were well aware of this. Diogenes Laertius writes: "The god Phoebus gave birth to two physicians for man: Aesculapius for the cure of the body, Plato for the cure of the soul" (*Lives of Eminent Philosophers*, III, 45). And this truly remains one of the greatest messages of Plato for man today: remember that if you want to free yourself from many of your evils, you must in the first place cure your soul.

Università Cattolica del Sacro Cuore
Milan, Italy

NOTE

[1]For a development of this point, see G. Reale (1995, passim).

BIBLIOGRAPHY

Diogenes: 1925, *Lives of Eminent Philosophers,* vols. 1 and 2, Loeb Classical Library, Cambridge, MA/London.
Plato: 1968a, *Charmides*, W.R.M. Lamb (trans.), Loeb Classical Library, Harvard University Press, Cambridge, Massachusetts.
—— 1968b, *Republic*, P. Shorey (trans.), Loeb Classical Library, Harvard University Press, Cambridge, Massachusetts.

—— 1968c, *Timaeus*, R.G. Burey (trans.), Loeb Classical Library, Harvard University Press, Cambridge, Massachusetts.

—— 1968d, *Apology*, H.N. Fowler (trans.), Loeb Classical Library, Harvard University Press, Cambridge, Massachusetts.

Reale, G.: 1995, *Per una nouva interpretazione di Platone. Rilettura della Metafisica die grande dialoghi alla luce delle dottrine non scritte*, Edizione Cusl, Milano.

PAULINA TABOADA

THE GENERAL SYSTEMS THEORY: AN ADEQUATE FRAMEWORK FOR A PERSONALIST CONCEPT OF HEALTH?

I. INTRODUCTION

The conviction that a reductionist view of the world and the exclusive application of the analytical method in science have not been able to give us either an adequate theoretical framework or good operational models for health and health care underlies most of the contemporary literature on the so-called 'health crisis 2000'.[1] A concern shared by many authors interested in this topic is the need to overcome the old reductionist models of health and to provide an approach that does justice to the human person *as a whole*.

In the last few decades, different approaches to the concept of human health have been proposed. A number of these 'new' definitions can be grouped under what I shall call here 'systems conceptions of health,' because they are grounded on the General Systems Theory. Such definitions intend precisely to do justice to the person as a whole, and to overcome the shortcomings of reductionist approaches to the human person in medicine. We find examples of this in the theories of authors such as Anderson (1984, 1988), Caplan (1981), Capra (1983), Dubos (1979), Engel (1960, 1977), Kaplun (1992), Lazlo (1972), Noack (1987, 1988, 1991), Purola (1972), Sobel (1979), and many others.

The task of this paper is to test whether General Systems Theory succeeds in providing a cogent theoretical framework for a truly personalist conception of health. By 'personalist' I understand a conception of human health that includes the category of 'person' in the basic characterization of what health is. Proceeding as if one could understand human health without reference to the personal character of the being whose health it is marks the fundamental difference between the various conceptions I have called reductionist above and any personalist conception, which is based on the fundamental idea that one cannot understand human health without reference to human personhood.

I shall argue that despite its valuable contributions and broad acceptance, a number of difficulties arise from the attempts to grasp the specifically personal dimensions of human health by applying the General

P. Taboada, K. Fedoryka Cuddeback and P. Donohue-White (eds.), Person, Society and Value: Towards a Personalist Concept of Health, 33–53.
© 2002 *Kluwer Academic Publishers. Printed in Great Britain.*

Systems Theory. Thus, some of the conceptions that appear to be non-reductionist turn out in the end to reduce the datum of health and the human person to a non-personal equilibrium of purely material factors. But already the irreducibility of life to matter proves the irreducibility of health to purely material functions. Moreover, there are also some difficulties with respect to the account of individuality and personhood given by the different systems conceptions of human health.

Therefore, I shall argue that understanding human health requires the introduction of new categories. Indeed, there are specifically human and personalist dimensions of health that need to be explicitly addressed in a conception of health that does full justice to the nature of the human person.

II. THE SYSTEMS CONCEPTION OF HUMAN HEALTH

Before entering into a discussion of the positive and negative aspects of the systems conception of human health, I find it necessary to present some basic principles of the General Systems Theory, especially those concerning more directly the understanding of the phenomenon of life and the nature of the human person, in which any definition of health has to be grounded.

A. The General Systems Theory

Conceived in the middle of this century, the General Systems Theory became popular during the seventies and continues to be widely accepted and applied in the various fields of science to this day (mainly in computer sciences, but also in social and political sciences as well as in theoretical biology and medicine). It can be considered as the present paradigm[2] in theory of science.

General Systems Theory understands itself as a "general science of wholeness" (Bertalanffy, 1968, p. 37). The principal aim of this theory is to attain, with a high level of abstraction, knowledge of the most general principles that explain the structural similarities ('isomorphisms') and the organization in different realms of being. Thus, it searches for an understanding of the unifying principle of reality.

The Austrian theoretical biologist Ludwig von Bertalanffy (1952, 1963, 1968, 1975) was one of the originators of this theory.[3] He concludes that "the unifying principle is that we find organization at all levels" (Bertalanffy, 1968, p. 49). Classical analytic method tends to dissect a given phenomenon in order to study the behavior of its different parts separately, afterward reconsidering the whole as the sum of the parts. Systems perspective instead proposes a synthetic approach to reality, focusing the attention on those phenomena that need to be understood as organized wholes. In other words, the General Systems Theory intends to attain an understanding "not only [of] parts and processes in isolation, but [of the] problems of order and organization unifying them, resulting from dynamic interaction of parts" (Bertalanffy, 1968, p. 31).

Within this conception, the world is seen as a series of hierarchically ordered systems continuously interacting with each other. A 'system' is defined as a "set of elements standing in interaction" or as an "organized whole" (Bertalanffy, 1968, p. 38). According to this perspective, there are different levels at which we find the phenomenon of 'organized wholeness', such as a cell, an organ, a living organism, a community, the society, the world or even the cosmos. Depending on which level seems to be more adequate for the study of the phenomenon we are focusing on, one defines this level as a system. The necessary and sufficient condition for defining a system at any level of being is to discover the 'principle of organization' unifying the different elements constituting it. Organization is thus conceived as the unifying principle, i.e., that which accounts for the unity of a system at any level.

Bertalanffy conceives of the principle of organization as the most general unifying principle of reality. But organization is in turn considered to be the result of the order of processes of interaction between the different parts or elements of a system. Thus, the unifying principle of reality is conceived in a dynamic way, i.e., in terms of dynamic relationships or processes of interaction. As a result, the concepts 'dynamism' and 'process' assume a central role within the systems perspective. Capra writes that "to think of reality in terms of systems means to think in terms of processes" (1983, p. 295).

According to this theory, the hierarchical order of systems (the levels of organization) cannot be arbitrarily changed; it is, so to speak, given. What actually changes is the perspective or point of view from which we look at reality, trying to come to an understanding of it. Nevertheless, the perspective we choose in each case is left up to us, according to the specific level of organization we are focusing on. This approach to reality is, therefore, called 'perspectivistic'.

In this hierarchical order of systems, each level of organization shows specific properties that can neither be reduced to nor derived from the properties of lower levels, but which arise from this specific level of organization as a whole. In other words, the new properties of higher levels of organization do not arise from the parts as such, but from the relations or dynamic interactions between them. The assertion that the "whole is more than the sum of its parts" and the recognition that this whole compared with its components has new properties and modes of action are two fundamental postulates of the General Systems Theory.

B. The Systems Conception of Life

Bertalanffy's application of the systems approach to problems of theoretical biology gave rise to the so called 'organismic conception of life'. To grasp the nature of life is perhaps one of the most difficult problems of the philosophy of nature. For the sake of brevity, I will roughly group the different historical attempts to account for the unique phenomenon of life into two main streams: the vitalistic and the mechanistic conceptions. Bertalanffy's theory was conceived as a 'third way' between these two classical alternatives.[4]

According to his organismic conception, living organisms are defined as self-organizing unities, characterized by their capacity to preserve a 'steady state' ('Fließgleichgewicht')[5] through a constant exchange of matter, energy and information with their environment. For Bertalanffy (1968, p. 158) this is the "fundamental mystery of life." In order to account for what he considered to be the essential mark of life, namely self-organization, this author found it necessary to introduce the notions of 'open systems' and 'steady state'. Since these notions are basic presuppositions of the present systems definitions of human health, I shall unfold their meaning here.

Bertalanffy's notion of 'open systems' intends to point out the essential difference between living beings and physical systems, which are by definition closed systems and which, according to the second law of classical thermodynamics, tend spontaneously to a state of maximal disorder. The peculiar capacity of living beings to preserve or produce order from disorder ('negentropy'), maintaining themselves as organized wholes rather than disintegrating, results, according to Bertalanffy, from the continuous exchange of matter, energy and information with the environment.

Thus, the defining feature of living systems according to this author is the fact that they are open, i.e., that they constantly incorporate new matter, energy and information, which allows them to preserve their order and organization in spite of the spontaneous tendency to disorder we encounter in nature. Thus, the above-mentioned key role that dynamic processes of interaction have for the General Systems Theory becomes even stronger in its particular application to living beings. "The basis of the open-system model is the dynamic interaction of its components" (Bertalanffy, 1968, p. 150).

On the other hand, according to Bertalanffy, precisely this 'openness' and the continuous processes of interaction account for a second essential mark of living beings, namely the capacity to preserve a 'steady state'. Bertalanffy's notion of 'Fließgleichgewicht' refers then to an essential feature of living systems, namely their unique ability to maintain a peculiar type of balance: a 'flowing balance'. This notion is based on the empirical demonstration that all the matter making up an organism is replaced within a specific period of time, which varies depending on the species and the particular organs and tissues.[6] Nevertheless, despite this constant renewal of their constitutive elements, living organisms are able to preserve their specific structural and functional unity. Bertalanffy's notion of 'steady state' refers precisely to this fact.

Thus, in the organismic conception of life, both the structural and the functional unity of a living system are explained in terms of dynamic processes. What constitutes the apparently lasting (static) structure of living organisms is not really static but a result of very slow processes of permanent loss and gain of matter and energy. On the other hand, the functional unity results from the order of faster processes of interaction.

C. The Systems Conception of Human Health

The results of applying the General Systems Theory to defining human health can be shown through examples of what I have called 'systems conception' of human health.[7] Some of the authors referred to above are very explicit in stating that the General Systems Theory is the theoretical framework for their definitions of health.[8] Others do not mention their theoretical background, but as soon as one becomes familiar with the basic concepts of the systems perspective, it becomes easy to recognize the common features resulting from its application to defining human health.

The common element of systems definitions of health is the fact that they stress two main aspects of this phenomenon: health as a 'dynamic balance' and health as an 'ability to cope'. Since according to the General Systems Theory the essential features of living beings are their being 'open systems' and their ability to maintain a 'steady state', both the conception of health as an ability to preserve a state of dynamic balance, and the idea that healthy living organisms have the potential to adapt to the always changing conditions of their environment in a meaningful and successful way result logically from the application of this theoretical framework to defining health. We find a clear example of this, for instance, in Noack's definition of health: "Health can be defined as a state of dynamic balance - or more appropriately as a process maintaining such a state - within any given subsystem, such as an organ, an individual, a social group or a community" (1987, pp. 14-15).

Hence, two key dimensions of health can be defined: health balance and health potential. While some authors tend to focus on the process of maintaining a state of dynamic balance, others stress more the ability to cope as the key mark of healthy organisms. The present tendency of the systems conceptions of health seems to place greater emphasis on this second aspect, conceiving of health basically as the ability of an individual or a given community to cope.[9]

Regardless, however, of whether a specific author considers one or the other of these two dimensions more relevant for defining health, a common insight underlies the systems definitions of health, namely, the dynamic nature of our datum: health is fundamentally conceived in terms of dynamisms. Noack again states: "Health is thus viewed as a dynamic characteristic of the individual, the social group or the socioecological system: it is clearly associated with the activities of such systems as a whole or of their component parts" (1987, p. 14).

D. The Systems Conception of the Human Person

An important insight of the systems conception of the human person is the idea that human behavior cannot be understood adequately by applying mechanistic criteria to interpret it. In fact, an explicit goal of the General Systems Theory is to overcome the shortcomings of former mechanistic conceptions of the human person and specifically human behavior. If a mechanistic understanding of the person explains human behavior as a reaction to different external stimuli (i.e., in terms of a linear cause-effect relationship), the systems conception stresses rather the person's original moment of creativity. "Man is not a passive receiver of stimuli coming from an external world, but in a very concrete sense creates his universe" (Bertalanffy, 1968, p. 194). For Bertalanffy, "a new model or image of man seems to be emerging. We may briefly characterize it as the model of man as active personality system" (1968, p. 192). According to this author, "emphasis on the creative side of human beings, on the importance of individual differences, on aspects that are non-utilitarian and beyond the biological values of subsistence and survival - this and more is implied in the model of the active organism" (Bertalanffy, 1968, p. 193).

The fact that the human person occupies a unique position in nature and that her behavior can be clearly distinguished from that of other living beings seems evident to Bertalanffy. As a theoretical biologist, he asks himself about the reasons for this. His conclusion is that the task of defining the human person and her specific behavior goes beyond any unilateral consideration of the natural sciences and should include also philosophical and even theological reflections. Different sciences complement each other in trying to attain a proper understanding of the person (Bertalanffy, 1963).

Despite this statement, Bertalanffy attempts an exclusively biological approach to defining the human person. He says that "the systems concept tries to bring the psychophysiological organisms as a whole into the focus of the scientific endeavor" (1968, p. 193). He tries to identify what a biologist from an exclusively empirical point of view would be able to recognize as the specific marks of human beings, that which sets the person apart from other living organisms, such as plants or animals. His conclusion "as a biologist who attempts to characterize the special position of man in the cosmos" is that

> the privilege of mankind, that in which his psychology and his behavior consists, is the fact that man creates a world of symbols and lives in it. ... this definition is in fact both sufficient and necessary to circumscribe human behavior, language, culture and history, from the exclusively biological field. ... That, which makes man precisely man is the world of symbols and the realm of human values (1963, pp. XXIII-XXIV[10]).

In other words, the result of Bertalanffy's scientific search for the essential features of human behavior is that "symbolism is recognized as

the unique criterion of man by biologists [...] The distinction of biological
and specific human values is that the former concerns the maintenance of
the individual and the survival of the species; the latter always concerns a
symbolic universe" (Bertalanffy, 1968, pp. 216-217).

Thus, this author admits the existence of a number of values shaping a
truly personal life. He writes, "Man has values which are more than
biological and transcend the sphere of the physical world" (1968, p. 197).
Hence, at the level of human beings Bertalanffy discovers completely
new properties and capacities, which set the human person apart from
other living beings.

According to the systems conception, what accounts for these new
properties of the person is the 'psychophysical whole' as such. In
Bertalanffy's words,

> the properties and modes of action of higher levels are not explicable
> by the summation of the properties and mode of action of their
> components taken in isolation. If, however, we know the ensemble of
> the components and the relations existing between them, then the
> higher levels are derivable from the components (1952, p. 148).

Furthermore, for Bertalanffy, that which in the last analysis accounts
for the unique capacities of the human person is "a brain which allows
consciousness that by means of a world of symbols grants foresight and
control of the future" (Bertalanffy 1952, p.108). We shall come back to
this point later (Section III).

A further relevant aspect of the systems conception of the human
person is its account of individuality. In spite of stating that "strictly
speaking, there is no biological individuality" (1952, p. 49), Bertalanffy
considers the human person "before and above all an individual" (1968,
p. 52). How does he justify this assertion? According to him, "an
individual can be defined as a centralized system" (1968, p. 71). With this
he intends to stress the fact that primitive living organisms can be easily
divided: they are not 'individuals' in the sense of 'indivisible'. In
contradistinction, higher organisms cannot be so easily divided.

Bertalanffy's explanation for this fact is that in higher organisms
certain parts or organs gain a leading role, determining the behavior of the
whole. In this way they become what he calls 'centralized organisms'.
According to Bertalanffy, "the principle of centralization is that of
progressive individualization" (1968, p. 71). This process of progressive
centralization and individualization culminates in the formation of the

highly developed brain of the human person, who therefore can be considered "before and above all, an individual" (Bertalanffy 1968, p. 52). The concept of 'individual' implicit in this conception will be discussed in the next section.

III. CRITICAL ANALYSIS OF THE SYSTEMS CONCEPTION OF HUMAN HEALTH

In the previous section an ideal conception of the human person as an individual open to meaningful interactions with her environment and able to shape her behavior according to values which transcend the merely physical world arose. Nevertheless, in spite of the explicit intention of Bertalanffy to provide a theoretical framework able to account for these essential features of the human person, I find it problematic or even impossible to ground his statements to this effect in the General Systems Theory itself. Moreover, I already consider the theory's account of life as insufficient or at least highly problematic. The task of this section is to point out these difficulties.

A. Philosophical Problems of the Systems Account of Personal Individuality

If we admit the theoretical presuppositions of the General Systems Theory, a number of inconsistencies arise in its attempt to ground personal individuality in an exclusively empirical understanding of the concept 'individual'. Following the reasoning of this theory to its logical conclusions, we have no ground for considering *unity* as a real property of beings. In the last analysis there is nothing like an ultimate *substrate* grounding the unity of a system. Each system is rather reduced to a series of relationships, namely, to an order of processes of interaction and exchange of matter, energy and information.

Thus, the dynamic understanding of the unifying principle of being and the perspectivistic approach to reality proposed by the General Systems Theory imply a relativization of any kind of boundaries in nature. Bertalanffy was aware of this logical implication of his theory:

Strictly speaking, spatial boundaries exist only in naive observation, and all boundaries are ultimately dynamic. One cannot exactly draw the boundaries of an atom, ... of a stone, ... or of an organism (continually exchanging matter with environment) (1968, p. 215).

Bertalanffy also draws the practical conclusion resulting from these theoretical presuppositions, namely the rejection of biological individuality: "strictly speaking, there is no biological individuality" (Bertalanffy 1952, p.49). The relativizing of boundaries in nature and the negation of biological individuality implied by his theory poses a tremendous problem for Bertalanffy's conception of the human person, because he cannot and does not want to deny personal individuality. In fact, the opposite is the case, as stated in the previous section.

Aware of this problem, Bertalanffy tried to solve it by introducing what he called the 'principle of centralization'. It is after introducing this principle that he modifies his previous radical negation of biological individuality, saying: "Strictly speaking, biological individuality does not exist, but only progressive individualization in evolution and development resulting from progressive centralization ... so that the organism becomes more unified and 'indivisible'" (1968, p. 73).

Starting with the general theoretical presuppositions of the General Systems Theory it is not possible to introduce individuality as a *real* property of being without falling into logical contradictions. The only consistent way in which individuality could be reintroduced if one remains true to the general postulates of this theory is by means of considering it as a projection of human mind into reality, i.e., as a suitable perspective to look at some phenomena we are trying to account for, but not as an *objective* property of being.

This epistemological implication of the General Systems Theory was not recognized by Bertalanffy himself, but becomes explicit in the later application and development of this theory, for instance, in Maturana (1972).[11] Therefore, also in the later developments of the theory I see a confirmation of my points of dissatisfaction with the General Systems Theory.

B. Philosophical Problems with the Systems Account of the Specifically Personal Capacities

The systems conception of the human person runs into further difficulties as soon as it tries to account for what Bertalanffy considered to be the specific mark of human behavior, namely creativity. As stated in the previous section, Bertalanffy conceives of creativity and the specific features of human behavior as properties emerging from the psychophysical whole as such, and, in the last analysis, as properties emerging from the activities of the human brain.

Since the theory does not admit any non-material principle constituting the psychophysical whole, we have to assume that the same elements constituting non-living systems also constitute the human person. Following the inner logic of the theory, it does not become clear why such a psychophysical whole should give rise to characteristics different from those of other physical wholes. There is no principle given according to which we can understand why organization in the case of human persons gives rise to something other than organization in the case of non-personal beings.

We are left with nothing more than the statement that personal organization is different from non-personal. In the end, the specific and unique features of human beings, such as the "ability to create a world of symbols and to live in it," psychic life, consciousness, knowledge, affective life, value-responses, etc., have not been explained. We have no foundation for Bertalanffy's claim that human persons cannot be understood in purely biological terms and therefore no cogent argument against reductionism. In this way Bertalanffy fails to achieve an explicit goal of his theory.

C. Philosophical Problems with the Systems Account of Life

The theoretical problems arising from the systems conception of the human person are already present in the systems attempt to account for the phenomenon of life as such. Bertalanffy had a clear insight into the irreducibility of life to other phenomena. He understood that the essential marks of life, such as self-organization, self-building, goal-directed behavior ('teleology'), etc., cannot be adequately explained in mechanistic terms. This was his reason for rejecting any form of mechanistic reductionism. However, he did not see the need to introduce a vital principle to account for self-organization and the other essential features of life. In his organismic conception, the processes of exchange of matter, energy and information with the environment account sufficiently for the specific self-organizing capacity of living organisms.

Nevertheless, if a living being is conceived as a set of interacting elements, whose peculiarity is the ability to maintain a steady state through a continuous exchange of matter, energy and information with the environment, then I do not see a sufficient reason for explaining why only certain kind of systems would have the ability of self-organization while others do not. In such a case, there would be no essential, recognizable difference between organic and inorganic systems. How could inanimate matter exchange matter, energy and information with its medium in such a way that it would be able to move from a lower to a higher form of organization? If there were nothing like a principle of life already present in a system, making of it precisely a *living* system, these meaningful interactions could not take place at all.

With his theory, Bertalanffy does actually describe an essential feature of life, namely the capacity of self-organization, but he does not provide an ultimate non-reductionist foundation for it. The claim that self-organization is the specific mark of living organisms remains without any cogent theoretical foundation. Therefore, the organismic conception of life fails to be a 'third way' between the two classical alternatives, namely vitalism and mechanistic reductionism.

IV. CHALLENGES TO THE SYSTEMS CONCEPTION OF HUMAN HEALTH

In spite of their lack of an adequate theoretical foundation, I do recognize a number of positive insights in the systems conceptions of human health. This confronts us with the challenge of providing a cogent philosophical foundation for them, in order to secure a truly personalist conception of health.

A. The Challenge of an Adequate Philosophical Explanation of Personal Individuality

The first positive contribution of the systems conception of human health is the fact that it confronts us with the necessity of overcoming reductionist models of health, focusing our attention on the human person as a whole. Indeed, the systems conception opens an interesting path for this search. Moreover, by stressing the importance of considering the human person not as an isolated being, but as a being open to meaningful interactions with other persons and with her environment, this theory encourages the consideration of the multiple elements and dimensions involved in the maintenance or loss of human health. Hence, it invites us to take into account biochemical factors, personal interactions, family atmosphere and even socioecological elements influencing the human person's health. Within this conception, an alteration at any of these different levels influences the whole as such.

In fact, the inclusion of the relational character of the person in the systems conception of human health has led to what has been called the 'socioecological paradigm' of health.[12] According to this model, the multiple relationships and interactions that a person has with other persons in her family and society, with her environment and even with the whole cosmos, must be considered as constitutive parts of human health. They should therefore also influence health care policies. This is certainly a positive development in the concept of health care.

There is, however, a danger in such a conception, namely that the concept of health expands in such way that it is no longer possible to define its bearer or determine any responsibility for health care. Hence, the inclusion of the relational dimension of the person in the conception of human health needs to be grounded in a clear distinction between the individual as such and everything else to which she stands in relation. Thus, the systems conception of human health confronts us with the challenge of providing an adequate philosophical foundation for the unifying principle of being, which secures both the individual nature of the human person (which stands at the foundation of her being an autonomous subject of rights and duties), and her relational character (which is certainly an essential mark of personal beings).

Nevertheless, as stated above, the theoretical account of personal individuality is not sufficiently secured in the systems approach to reality. Moreover, the theoretical presuppositions of the General Systems Theory necessarily end up dissolving the very notion of individual, replacing it with the idea of a set of complex interactions, as we already saw. Therefore, I would recommend complementing this approach with an understanding of individuality as a specific type of intelligible unity or essence of a being, rather than conceiving of it as mere empirical indivisibility.[13]

Classical philosophy solved the problem of accounting for the individual nature of the human person through the idea of substance.[14] But regardless of any metaphysical commitments and the concrete solution I would suggest in order to solve this difficulty of the systems conception of individuality, it seems quite evident that the General Systems Theory requires some kind of revision in order to account sufficiently for personal individuality as a real property of being.

B. The Challenge of an Adequate Understanding of the Dynamic Structure of the Human Person

A second positive element of the systems conception of human health is its insight into the dynamic nature of this phenomenon, expressed in the recognition that living beings are endowed with intrinsic dynamisms tending to preserve their specific structural and functional unity. I regard as a valuable contribution of this theory the fact that it directs our attention precisely toward these inner dynamisms in order to grasp the nature of health. Any definition of health presupposes an understanding of the nature of life, and life seems to be essentially related to movement or, better said, to self-movement.

However, the result of my analysis of the systems account of self-organization is that it fails in providing an ultimate non-reductionist foundation. Moreover, the General Systems Theory does not really account for the peculiar inner dynamic structure proper to living beings. What the theory actually does is describe this property of life in terms of mathematical formulas. As Conrad-Martius (1964) states, there is no way left to explain the transphysical phenomena of life other than mathematical symbolism, if one does not want to revert to mechanistic analogies on the one hand and does not admit metaphysical categories on the other.[15]

Therefore, a further challenge posed by the systems conception of health is to account for the dynamic structure of the human person in a non-reductionist way. For this search we will need to start with an in-depth understanding of the peculiar type of self-movement proper to living beings. And since according to the principle of causality every effect has to be proportionate to its cause, we will also need to ask ourselves about the sufficient reason for this unique type of self-movement.

The insight into the inner dynamisms proper to life and the need to give a sufficient reason for them lead Aristotle (De Anima, 402b ff.) to introduce the notion of 'entelechia'.[16] Thus, he characterized living beings as those beings (ens), which have their end (telos) within themselves. In other words, living beings are able to move themselves from within, towards the fulfillment of their end. In order to account for this inner dynamic structure of living beings, Aristotle considered it necessary to introduce the idea of a 'vital principle'.

Accordingly, an understanding of life-dynamisms in terms of a potency-act structure of being and the acceptance of an intrinsic causal principle moving living beings toward their full actualization might be a necessary complement to the systems conception of life and health. Moreover, since we intend to secure a truly personalist conception of human health, an adequate understanding of the specific dynamic structure of the human person might also be required.[17]

Since the experience of a healthy person reveals a number of physiological processes happening spontaneously in her body (most of them even unnoticed by the person) and also a richness of psychological, affective, cognitive and spiritual potentialities that can be freely actualized, I would suggest that we conceive human health as a condition in which there is no immanent impairment of the unfolding and actualization of these various kinds of dynamisms. Nevertheless, I have to specify that by considering human health as a condition in which these different potentialities of the person can be actualized, I do not intend to say that all of them must in fact be actualized in order to call a person healthy. Certainly, a full actualization of all our potentialities is not possible, because of the many limitations proper to the human condition.

A philosophically cogent conception of human health presupposes then an understanding of the specific dynamic structure of the human person and the directedness of her unique inner dynamisms toward specific values that represent objective goods for her nature.[18] In other words, a truly personalist concept of human health has to be grounded in an adequate philosophical anthropology. And one has to admit that the theoretical presuppositions of the General Systems Theory do not allow for a non-reductionist account of the dynamic structure of the human person, as has been already proven.

C. The Challenge of Accounting for the Unique Capacities of the Human Person

Closely related to the above-mentioned aspect is a further contribution of the systems conception of human health, namely the fact that this theory admits the existence of completely new properties at the level of personal beings, which set the human person apart from other living organisms. According to this view, the specific personal 'moment' of health is the capacity of the human person to deal with the changes occurring in herself and her environment in a meaningful and creative way, solving problems, transforming the world through her work, communicating with her fellows through language, creating in art and culture, etc.

Nevertheless, as stated above, the theory explains these specifically personal capacities as emergent properties of the psychophysical whole and does not admit the existence of a non-material principle constituting the human person. But if one accepts the evidence of the classical principle that the effect has to be adequate to the cause, one would have to admit that it is only possible to explain the specific capacities of the person (such as consciousness, knowledge, love, freedom, value-responses, language, creativity, etc.) in terms of a non-material principle. As Conrad-Martius (1949, 1964a) observes, it might be possible to conceive of these properties as emerging from the whole as such, but only if we already understand the notion 'whole' as referring to the essence of personhood. Therefore, a further challenge posed here to the systems conception of human health is the need for a theoretical explanation which would allow us to go beyond reductionist explanations of the origin ('efficient cause') of the specifically human properties of the person.

V. CONCLUDING REMARKS

The present study of the results of applying the General Systems Theory to defining the human health leads me to the conclusion that this theory fails in providing a cogent theoretical framework for a truly personalist conception of health. Indeed, the General Systems Theory does not adequately account for the data it seeks to explain. This failure is deeply rooted in the general epistemological presuppositions of the theory, which make it impossible to ground a definitive rejection of reductionism.[19] If the theory is unable to explain the very data to which it seeks to render justice, then I find it difficult to consider it an adequate theoretical framework for understanding the health of the human person.

Nevertheless, in spite of their lack of an adequate theoretical foundation, I do recognize a number of positive insights in the systems conceptions of human health. Rescuing these contributions would entail re-thinking some elements of the General Systems Theory, for the sake of preserving the original intent of the theory itself, and of advancing further in the achievement of its original goal. Thus, I consider it a positive result of this critical approach that it brought us closer to a philosophical account of these valuable insights of the theory, and to a more cogent foundation for a properly personalist concept of human health.

Pontificia Universidad Católica de Chile
Santiago, Chile

NOTES

[1] For a more detailed account of what is presently meant by 'health crisis 2000' see, among others, the contributions to this topic in: Pan American Health Organization (PAHO), 1986, 1992; O'Neill, P.D., 1982; San Martin, H., 1982.

[2] In this paper I shall use the concept 'paradigm' in the Kuhnian sense (Kuhn, 1962), although I do not agree with the epistemological assumptions that underlie the understanding of 'paradigms' and 'paradigm change' in contemporary theory of science.

[3] Other initiators of this theory are, for instance, Boulding, K., Rapoport R. and Gerard R. (cf. Bertalanffy, 1968, p. 15 ff.)

[4] Although both the vitalist and the mechanistic conceptions of life have good arguments in their favor, both approaches have been strongly criticized for different reasons. For example, the main argument against the vitalist conception is that it finds it necessary to introduce a soul-like factor in order to account for the capacities of self-organization, self-building and goal-directed behavior characterizing living organisms. The way in which such a soul-like factor interacts with matter still remains unclear for many thinkers such as, for instance, Bertalanffy himself. He therefore accuses this position of being unscientific: a form of 'animism'. On the other hand, merely mechanistic interpretations of life processes in terms of linear cause-effect relationships are unable to account for the complex phenomena of integrated wholeness, organization, goal-directed functioning, etc., which are central features of living organisms. To define life in terms of mechanic organization is to explain it away. Thus, Bertalanffy finds it necessary to propose a 'third way' between the two traditional alternatives: his 'organismic conception'.

[5] Bertalanffy adopts the English expression 'steady state', which was already introduced in theoretical biology to refer to the specific type of balance one encounters in the physico-chemical processes occurring in living organisms. He translated this expression into German as 'Fließgleichgewicht'. In my opinion, the German expression 'Fließgleichgewicht' reflects more accurately what the author is trying to point out here, namely a kind of 'flowing balance'.

[6] For a detailed unfolding of this point see Bertalanffy, 1968, p. 120 ff.

[7] In recent decades a number of these definitions of health have been proposed. See the Introduction for a list of the most representative authors working with systems conceptions of health.

[8] A good example of this is Noack, who considers the systems perspective as presented by Miller (1978) and Capra (1983) the most suitable for defining health and health care policies: "It is necessary to introduce briefly the general theoretical framework in which the author has chosen to formulate this proposal, namely the systems perspective. This perspective seems to be the most suitable because it fits the socioecological paradigm of health and accounts for

one of its central meanings, dynamic balance or equilibrium", and, "with regard to human health and disease, it is useful to see individuals as part of social units such as families or other primary social groups which in turn form subsystems of larger sociocultural, economic and political systems embedded into an even more global societal and ecological system" (Noack, 1987, p.14).

[9] Noack (1987, p.14), for instance, stresses this aspect of health: "From a system's point of view, coping potential is thus an important health resource and therefore an aspect of health." Another example of this is Kaplun (1992, p. 415): "The experience, the thinking, the research reported offer a stimulating view of health, expressed in terms of the ability of people to cope with their problems and change their environments."

[10] This and some of the following statements are contained in the preface to the Spanish translation (Bertalanffy, 1963) of Bertalanffy's 1952. There is no official English translation of this preface. Thus, I introduce here my own English translation.

[11] On the basis of this implication, Maturana (1972) goes so far as to propose to eliminate the notions of 'individuality' and 'teleonomy' which have played a central role in the various attempts to grasp the phenomena of life throughout history.

[12] Noack, for instance, states : "a multifactorial, ... socioecological paradigm is a suitable framework for explaining the conditions and causes of health and ill health and for guiding health-related activities as well as health and social policy" (1987, p. 10).

[13] A clarification of the philosophical meaning of the unifying principle of being and the unfolding of the various types of unity we discover in reality, as has been done for instance, by Hildebrand (1991) and Seifert (1977), would help in solving the theoretical inconsistencies raised by the dynamic and overly empirical conception of the unifying principle of reality proposed by the General Systems Theory. An accurate exposition of the problems raised by the epistemological presuppositions of the General Systems Theory and the way in which an adequate philosophical foundation could help in solving the theoretical problems arising from its conception of the unifying principle of reality, goes beyond the scope of this short paper. I have taken steps toward this in my (1996).

[14] Moreover, at the level of personal beings, the traditional Aristotelian notion of substance reaches its fullest and highest meaning, as has been shown by Seifert (1989, 1989a). It is not possible to unfold this deep philosophical insight in this short paper. For further development of this point see my (1996) and the other works cited in this essay.

[15] This quotation is not literal, but a free translation of the original German text. Cf. Conrad-Martius 1964, p. 160.

[16] For an analysis of the Aristotelian notion of 'entelechia' and its application in contemporary theoretical biology, I refer the reader to Conrad-Martius 1949, 1964, 1964a.

[17] Here I understand the expression 'dynamic structure of the person' as Wojtyla (1979) did in his book *The Acting Person*. Starting from an analysis of what he calls the 'experience of man', Wojtyla reaches an understanding of different 'dynamisms' in the human person, which he summarizes introducing the expression 'dynamic structure of the person'. I think that the deep insights about the nature of the human person and the different dynamisms we discover in her presented in this book, provide rich material to ground a personalist conception of human health. This is a point I have unfolded in my (1996).

[18] The concept of 'objective good' in relation to the notion of health has been unfolded in the paper of Donohue-White and Fedoryka Cuddeback included in this volume.

[19] I further unfold this point in my (1996).

BIBLIOGRAPHY

Anderson, R.: 1984, 'Health promotion: An overview,' *European Monographs in Health Education Research*, 6, 1-126. WHO Regional Office for Europe, Copenhagen.

Anderson, R.: 1988, 'A discussion document on the concept and principles of health promotion', in WHO (ed.), *Health Behavior Research and Health Promotion*. Oxford University Press, Oxford.

Bertalanffy, L.v.: 1952, *Problems of Life. An Evaluation of Modern Biological and Scientific Thought*. Watts and Co., Ltd. New York.

—— 1963, *Concepción Biológica del Cosmos*, Faustino Cordón (trans.). Ed. Universidad de Chile, Santiago de Chile.

—— 1968, *General System Theory. Foundations, Development, Applications*, George Braziller, New York.

—— 1975, *Perspectives on General System Theory, Scientific-Philosophical Studies*, George Braziller, New York.

Brody, H. (ed.): 1979, *Ways of Health. Holistic Approaches to Ancient and Contemporary Medicine*, Harcourt Brace Jovanovich, New York.

Caplan, A. et al.: 1981, *Concepts of Health and Disease. Interdisciplinary Perspectives*. Addison-Wesley, London.

Capra, F.: 1983, *Wendezeit. Bausteine für ein neues Weltbild*. (Aus dem Amerikanischen Übersetzt von Erwin Schuhmacher). Scherz, Bern.

Conrad-Martius, H.: 1949, 'Die schöpferische Entwicklung des Lebendigen', in H. Conrad-Martius (ed.), *Bios und Psyche*, Claassen & Goverts, Hamburg.

—— 1963, 'Die Seele der Pflanze', in H. Conrad-Martius (ed.), *Schriften zur Philosophie*, Vol. I. Kösel, München.

—— 1964, 'Präformismus in der Natur', in H. Conrad-Martius (ed.), *Schriften zur Philosophie*, Vol. II. Kösel, München.

—— 1964a, 'Zum Wesensunterschied zwischen Lebendigem und Unlebendigem', in H. Conrad-Martius (ed.),*Schriften zur Philosophie*, Vol. II. Kösel, München.

Dubos, R.: 1979, 'Medicine evolving', in H. Brody and S. Sobel (eds.), *Ways of Health. Holistic Approaches to Ancient and Contemporary Medicine*, Harcourt Brace Jovanovich, New York.

Engel, G.L.: 1960, 'A unified concept of health and disease,' *Perspectives in Biology and Medicine*, 3, 459-485.

—— 1977, 'The need for a new medical model: A challenge for biomedicine,' *Science*, 196(4286), 129-136.

Engelhardt, H.T.Jr.: 1981, 'The concepts of health and disease,' in A. Caplan et al. (eds.), *Concepts of Health and Disease. Interdisciplinary Perspectives*, Addison Wesley, Reading, Massachusetts.

—— 1996, *The Foundations of Bioethics* (2nd ed.), Oxford University Press, New York.

Hildebrand, D.v.: 1991, *What is Philosophy?* Routledge, New York.

Kaplun, A.: 1992, 'A dynamic vision of health,' in A. Kaplun (ed.), *Health Promotion and Chronic Illness. Discovering a New Quality of Health*, WHO Regional Publications, European Series n. 44, Copenhagen.

Kuhn, T.: 1962, *The Structure of Scientific Revolutions*, University of Chicago Press, Chicago.

Laszlo, E.: 1972, *The Systems View of the World. The Natural Philosophy of the New Developments in the Sciences*, George Braziller, New York.

Maturana, H. and Varela, F.: 1972, *De Máquinas y Seres Vivos*, Ed. Universitaria, Santiago de Chile.
—— 1984, *El Árbol del Conocimiento*, Ed. Universitaria, Santiago de Chile.
Miller, J.: 1978, *Living Systems*, Mc Graw-Hill, New York.
Noack, H.: 1987, 'Concepts of health and health promotion,' in T. Abelin, Z.J. Brzezinski, and V. Carstairs (eds.), *Measurement in Health Promotion and Protection*, WHO Regional Office for Europe, European Series n. 22, Copenhagen.
—— 1988, 'Measuring health behavior and health: Towards new health promotion indicators,' *Health Promotion*, 3 (1): 5-11.
—— 1991, 'Conceptualizing and measuring health,' in B. Badura, and I. Kickbusch, *Health Promotion Research. Towards a New Social Epidemiology*, WHO Regional Publications, European Series n. 37, Copenhagen.
O'Neill, P.D.: 1982, *Health Crisis 2000*, Regional Office for Europe of the WHO, Copenhagen.
Pan American Health Organization (PAHO): 1986, *Evaluación de la Estrategia de Salud para Todos en el Año 2000*,' in PAHO (ed.) *Séptimo Informe sobre la Situación Sanitaria Mundial*, PAHO, Washington, D.C.
—— 1992, *The Crisis of Public Health. Reflections for the Debate*, Scientific Publication n. 540, PAHO, Washington, D.C.
Purola, T.: 1972, 'A systems approach to health and health policy,' *Medical Care*, 10(5).
San Martin, H.: 1982, *La Crisis Mundial de la Salud. Problemas actuales de epidemiología social*, Karpos, Madrid.
Schippergers, H.: 1985, *Der Garten der Gesundheit. Medizin im Mittelalter*, Artemis, München-Zürich.
Seifert, J.: 1977, 'Essence and existence. A new foundation of classical metaphysics on the basis of "phenomenological realism", and a critical investigation of "existentialist Thomism",' parts I. and II. , *Aletheia* 1(1), 17-157; *Aletheia* 1(2), 371-459.
—— 1989, *Das Leib-Seele-Problem und die gegenwärtige philosophische Diskussion. Eine systematisch-kritische Analyse*, Wissenschaftliche Buchgesellschaft, Darmstadt.
—— 1989a, *Essere e Persona. Verso una Fondatione Fenomenologica di una Metafisica Classica e Personalistica*, Vita e Pensiero, Milano.
Sheldon, A.; Baker,F. and McLaughlin, C.: 1970, *Systems and Medical Care*,The MIT Press, London.
Sobel, S.: 1979, 'A systems view of health and disease,' in H. Brody and S. Sobel, (eds.) *Ways of Health. Holistic Approaches to Ancient and Contemporary Medicine*, Harcourt Brace Jovanivich, New York.
Spijk, P.v.: 1991, *Definition und Beschreibung der Gesundheit. Ein medizinhistorischer Überblick*, Schriftenreihe der Schweizerischen Gesellschaft für Gesundheitspolitik, Bd. 22, Muri.
Taboada, P.: 1996, *Systemtheorie und Gesundheitsbegriff. Kritik eines Reduktionismus im Lichte der Realistischen Phaenomenologie*. (unpublished Master's thesis, International Academy of Philosophy in the Principality of Liechtenstein).
World Health Organization (WHO): 1946, 'Minutes of the technical preparatory committee for the international health conference,' held in Paris from 18 March to 5 April 1946. *Off. Rec. Wld Hlth Org.*, 1, 1-79.
—— 1946a, *Proceedings and Final Acts of the International Health Conference*, held in New Jork from 19 June to 22 Juliy 1946. *Off. Rec. Wld Hlth Org.*, 2, 1-143.
—— 1983, *Constitution*. Reprinted from Basic Documents (37. Ed.), WHO, Geneva.

—— 1983a, *'The Health Centre Concept in Primary Health Care,'* *Public Health in Europe*, 22, WHO, Geneva.

—— 1984, 'National health systems and their reorientation towards health for all. Guidance for policy-making,' *Public Health Papers*, WHO, Geneva

—— 1986, *Advisory Committee on Health Research. Health Research Strategy for Health for All by the Year 2000*, WHO, Geneva.

—— 1992, *Our Planet, Our Health: Rapport of the WHO Commission on Health and Environment*, WHO, Genèva.

Wojtyla, K.: 1979, *The Acting Person*, Reidel, Boston.

PASCAL IDE

HEALTH: TWO IDOLATRIES[1]

> Insofar as medicine is concerned, it would not be good to reject this gift of
> God [i.e. the medical science] merely because of the bad use some
> people make of it. [...] On the contrary, we should bring to light the
> extent to which these people are corrupted.
>
> Saint Basil the Great, *The Longer Monastic Rules* (55, 3, PL 31, 1048).

It has become classical to oppose two conceptions of health, in relation to two types of medical practices, or even two approaches to the human body. Simplifying, I shall call the first conception *objectivist* and the second one *naturalist,*[2] epithets that will become clear later. The different terms available in English to designate a state of unhealth allow me to give a hint at this distinction. The English language has three different terms whose meanings are quite distinct and which can be contrasted with the poverty of French's single term *maladie.*[3] In English, *disease* refers to the malady as it is apprehended by medicine; *illness* signifies the malady as experienced by the patient; *sickness* suggests a state of malaise rather than of malady or, more precisely, a social representation of the malady, if one follows Jean Benoist's analysis of the American contributions to the anthropology of health and disease.[4]

The first model of health is linked to an objective conception of disease (disease-object), the second one to subjective (disease-subject) and social (disease-society) conceptions. Does such an opposition make sense? Do we have to adopt one of these two conceptions or, in a borderline case, can we have a mix of both?

I. THE TWO CLASSICAL MODELS OF HEALTH AND DISEASE

One needs to have a clear understanding of the fundamental distinction underlying the two classical models of health and disease in order to understand my argument. Hence, I shall clarify this distinction, propose three arguments that confirm it, and offer a path towards overcoming some of its problematic aspects.

P. Taboada, K. Fedoryka Cuddeback and P. Donohue-White (eds.), Person, Society and Value:
Towards a Personalist Concept of Health, 55–85.
© 2002 *Kluwer Academic Publishers. Printed in Great Britain.*

A. The Objectivist Model of Health

Two points need to be clarified from the very beginning. On the one hand, even though the biomedical model that shall be described here is currently dominant, it is neither taught nor reflected upon as such, *in actu signato*. It is rather part of a common unconscious representation conveyed through medical studies and practiced by medical staff, especially those working in hospitals, and remains the implicit point of reference of medical teaching and practice. On the other hand, the predominance of this model does not imply exclusivity. As I shall argue below, patients as well as doctors transgress this model all the time, a fact that shows its insufficiency *in actu exercito* and calls for another approach to health.

According to the model we are dealing with, health is an observable and quantifiable characteristic of the human body. It is reducible to a certain number of biological parameters. For instance, a healthy body has between 4 and 5 million red blood cells (half a million less for women), a potassium level between 3.5 and 4.5 milliequivalents per liter, and a weight (W) in kilograms such that the result of the fraction W/S^2 will be between 18.5 and 25 (where S indicates the size measured in meters). With values below the given range, we consider the body to be too thin; values above it indicate excess of weight; and an index above 30 points toward obesity, with the ineluctable consequences it implies.

In other words, according to this model, health presents the following four characteristics:

1) It is *objective*, in the sense that it refers to external data and not to subjective feelings. Consequently, the criteria of health can be universalized and normalized since they are not linked to individual representations. The American anthropologist Byron Good, who at the beginning of the 1990s was called to evaluate Harvard's medical programs and who met personally with students for several weeks, reports the following remark by a medical resident: "We are not here simply to make people tell their lives. As professionals, we have learned how to transcribe phenomenological descriptions of behavior into physiological and pathophysiological processes" (Good, 1998, pp. 174-175).

2) It considers the body according to the (macroscopic and microscopic) morphology of its organs and the physiology of its different functions. Added to the first characteristic, this second one signifies that the

health of the organs and of their functions is measured through *quantifiable*, statistical criteria, which can be normalized and regarded as the object of science. The body, Didier Sicard notes, "is more and more absent from medicine, and is present only if medicine offers it objective parameters, images or numbers" (1999, p. 157).

3) It is *somatic*, which means that it considers only the body, independently of the external environment and the psyche.

4) It is *invariable*, i.e., independent of the history of the subject and, therefore, independent of any eventual responsibility (cf. Genard, 1999). The biomedical model of health considers these parameters at a given time and is therefore able to say whether or not the body is healthy; independent of its future evolution. If I observe a blood sugar level of one gram between meals, I can certify that it is normal and then healthy.

Since the first of these characteristics commands the other three, I have labeled this conception of health "objectivist," the suffix stressing the partiality of such a perspective. This point has been well summarized by Leriche's famous saying: "health is life in the silence of organs". The objectivist model of health is based on a vision of the body that I would sum up in two features: *passive* and *fragmented*.[5] As a matter of fact, the first characteristic of a body that can be objectivized is its quantification; what Descartes would have named its extension. But extension is a passive characteristic; activity is a qualitative property. Therefore, to stress measurability is to make oneself blind to the dynamisms supporting the human body. On the other hand, what is extended is dividable, decomposable into parts.

Here, the requirements of the scientific approach converge with those of biotechnique. Indeed, technical manipulation requires fragmentation.[6] This is particularly clear for two new techniques, namely the transplantation of organs[7] and the sequencing of the genome.[8] This is the reason why the objectivist model and the anthropology of the passive body lead to maximizing the place of technique and minimizing the role played by the organism.

The most evident consequence of this objectivist approach is to dispossess the subject of his pathology and those who look after him of their compassion. "Medicine formulates the human body and the disease in a culturally distinct way" writes Byron Good (1998, p. 150). Anne Fagot-Largeault, medical doctor and philosopher, speaks of a "tradition of hardening. As if the contact with human suffering was creating a compensatory need to keep oneself at a distance, or even to laugh the suffering off" (see an interview of Fagot-Largeault in Canto-Sperber, 1998, p. 102).

While psychologists detect here a self-protective mechanism designed to cope with an omnipresent culpability (cf. Naquet, 1998, pp. 57-69), philosophers decipher it as an encounter between an anthropology of the objectified body and an objectifying look that abstracts the humanity out of the subject. I shall exemplify this point with a personal anecdote. During my residency in nephrology at Tenon, we admitted a patient with severe renal insufficiency due to septic shock; he had been saved *in extremis*. Finally, on the day he left the hospital, we went to his room. We were visiting the patients that day; significantly, almost all of the previous "medical visits" took place in the non-resident medical student room, with only the medical chart. The patient, deeply conscious of the fact that he had escaped death thanks to the technical skill and zeal of the medical staff, began warmly to thank the attending physician. Tears even came to the patient's eyes. The attending physician quickly left the room, mumbling incomprehensible words. A bit more and he would have asked the nurse to prescribe a lachrymal dryer for the patient! I had the impression that this doctor was, for the first time, meeting a sick person, and not merely a disease.

B. The Naturalist Model of Health

Holistic approaches to health did blossom during the first decades of the 20th century. In order to maintain health or promote recovery, some physicians emphasized the role of nutrition, thermal cures, stays at the mountain, and sanitaria that offered the advantages of rest, pure air and sun. Along with these approaches, the so-called 'natural remedies' also became very popular (i.e., all the remedies that can be obtained from traditional medicinal plants). During the 1920s, naturopathy developed thanks to a polygraphic doctor, Paul Carton.[9] Supported by a national association and a journal, this synthetic doctrine is based on a semi-vegetarian diet and some physical and mental exercises, all tinctured with hermeticism.

The evolution of nouns is interesting. For three-quarters of a century, different medical approaches, among which homeopathy is the most famous, were ignored and marginalized by official medicine. Later, during the 1970s, they were termed *alternative* medicines. It was a *de facto* acknowledgment of their specificity, but in a dialectic and discriminatory way; the epithet *alternative* implies a mutual exclusion, in such a way that to accept conventional medicine automatically means to disqualify the other forms, and *vice versa*. At the beginning of the 1980s, the less polemical term *complementary* therapy was proposed. Today, we speak of *integrated* therapy, an adjective that expands the field of the therapies we used to consider appropriate only for functional patients (sometimes called also "mild medicine") to the more serious problems of contemporary medicine.[10]

Another model, defining health as an inner state of harmony of the body in relation to its environment, its psyche, and its history, has also been contrasted with the objectivist model. Such a conception of health is opposed point by point to the four properties of the objectivist model described above:

1) Health is here a *subjective* characteristic: to *be* healthy is to *feel* healthy, not only in the negative sense of not experiencing anything disturbing, but also in the full, positive sense of feeling well, peaceful and ready to act at the same time, with a body that answers to what is asked without complaining. Health here takes the body-subject into account, i.e., the actual experiences of the body.

2) It corresponds to a *quality* of life that can neither be reduced nor analyzed in parameters objectifiable. To use Canguilhem's categories (1966, 1992), health does not belong to "normality" (which is quantifiable) but to normativity, that is to say, to value. Does the Latin *valere* not testify to the deep relationship between health and value?

3) Health is an *entering into relation* with the other of the body, still internal to the person, namely the psyche, as well as with the external other of this other, namely the environment and society.[11] Thus, health becomes not *normality* but rather *norm*: capacity to cope with the environment.

4) It is *historical*. The health of a person is inscribed within a history. Even though after the treatment and convalescence the patient might recover *ad integrum* the health he used to have, the pathology is part of his memory and his experience will not leave him uninjured.

Analogously to the way in which I characterized the objectivist conception of health above, this new vision of a healthy being can be described as based on a representation of the body, here *active* and *unified*. If health is a value, it is because it is a capacity of the body. The insistence on the dynamics proper to the organism leads to minimizing or even fearing and devaluing biomedical technique, and trusting more in the unlimited capacities of recovery of the body. This is the reason why such a vision of health has been called *naturalist*. Furthermore, it considers the human body in its entirety. Far from being linked to the state of one organ or one function only, health becomes a holistic property.

We see that each of the two models of health grows out of its contrary, that is to say, from what the other conception excludes. From a philosophical point of view, such an opposition is actually ideal, as most of the time we face models and practices that are hybrids.

C. Three Confirmations of these Different Approaches

The different approaches to health described above find a confirmation their opposite (i.e., in the conceptions of disease), as well as in the history of medicine.

1. The two views of disease

In order to grasp the elementary forms of disease and recovery, I shall refer to the remarkable work of François Laplantine (1986).[12] In the same manner in which two conflicting views of *health* have been proposed, two opposing approaches to *disease* have been suggested. According to Laplantine (1986), this distinction can be explained through four main categorical pairs that I have summarized in the following table.[13]

First approach	Second approach
Ontological model	Relational (or functional) model
Exogenous model	Endogenous model
Additive model	Subtractive model
Malefic model	Benefic model

According to the ontological model, medicine is centered on disease, i.e., on an ontological category. The relational model focuses instead on the sick person, i.e., on a functional level. At first sight, this bipolarity (ontological-functional) seems to command the other distinctions (Laplantine, 1986, p. 76). In reality, a careful reading of the book shows that the decisive distinction is the second one, i.e., the distinction between the exogenous model (according to which the cause of the disease is external) and the endogenous model (according to which the cause is internal). This distinction allows us to explain the other elementary forms of disease and recovery as well. We have already seen, and we shall see again, that the objectivist representation of health is different from the naturalist one inasmuch as the latter gives primacy to the (internal) activity of the body helped by medical technique and the former to the (external) activity of the same technique making up for the passivity of the body.[14]

2. The history of these two conceptions of medicine
The history of medicine is often presented in two or even three main stages. At the beginning is a mainly magical, pre-rational medical practice. Later, with Hippocrates, and even more so in the Renaissance with Andries Van Wesel, a scientific medicine appears that will show its full measure only in the 1880s (according to the usual chronology). A few observers add a third period, in which a vision centered on the patient appears as a reaction to the objectifying practice of medicine centered on disease.

The vision of a history of medicine in two stages (pre-rational and then progressively more reason-based), seems to be victim of a positivist philosophy of history and naively ignores the irrational part and the non-scientific presuppositions (coming, for instance, from scientific postulates) in contemporary medical practices. Moreover, even though the vision in three stages honors the appearance (some people would say the reappearance) of functional medicine, this conception is also disputable. As a matter of fact, can we not find *from the very beginning* in ancient cultures *a juxtaposition* of at least two types of practices of medicine or of functions of doctors, if not already two types of doctors (a more magical and a more empirical one) without a perfectly clear line of demarcation? Indeed, man has always striven to interpret coherently the real world in order to tame a cruel environment and to give meaning to the hardship of life.[15] Furthermore, he has had a double approach to it: one more immediate and empirical that consists in analyzing symptoms and causes, the other more global, determining their meaning and goals in the midst of life, nature, and society. Disease is actually one of these existential ordeals.

In this way, in Mesopotamia two types of medicine met in practice, if not in their conclusions: a "medicine of doctors" and a "medicine of magi" (Bottero, 1984). The paths are different: the doctor diagnoses what the ailment is and uses drugs that he chooses and prepares. The magus does not have such freedom. Moreover, he rather tries to inscribe the ailment of the disease within the field of existence, of the universe built by myth, where demons, intermediary entities between gods and human beings, interfere. Finally, he uses, according to certain laws (of similarity or contrariety), two essential means of conjuration, namely manipulation and speech.[16]

In the modern occidental era, with the birth of clinics, the personhood of the sick person and his suffering began to be forgotten, affirms Foucault (1975b). At the hospital, the subject is first and the disease second; the contrary happens at the clinic where the sick person becomes ultimately an object.

At the hospital, we deal with individuals who indifferently bear one disease or another; the role of the doctor is to discover the disease in the sick person; and this internality of the disease is such that it is often buried in the sick person, hidden like a cryptogram. At the clinic we deal on the contrary with diseases whose bearer is indifferent: what is present is the disease itself, in the body that is proper to it and which is not the body of the sick person, but the body of its truth. The sick person is only that through which a sometimes complicated and confused text is given to be read. At the hospital, the sick person is the subject of his disease; that is to say, we face a case; at the clinic, where we deal only with examples, the sick person is the accident of his disease, the transitory object that the disease appropriated to itself (Foucault, 1975b, p. 59).

Each of the two perspectives – synchronic and diachronic – probably entails part of the truth, that which can be explained through anthropological mechanisms such as pendular reaction, compensations, and the limits inherent to any systematic perspective. It does not really matter. In any case, we are facing two visions of health that often confront each other, sometimes managing a compromise, but never merging.

The preceding considerations can be summarized in the following table:

	Health according to the objectivist model	Health according to the naturalist model
Representation of *health*	Objective, quantifiable, somatic, a-historical	Subjective, qualitative, global, historical
Representation of the *body*	Passive and fragmented body	Active and unified body
Place of *medical technique*	Over-valorization of technique	Under-valorization or denial of technique
Representation of the *disease*	Ontological, exogenous, additive and malefic model	Relational (or functional), endogenous, subtractive and benefic model

D. A Criticism of the Bipartition

Is this dichotomy as clear as it looks? There are three main reasons for doubting it. First, these two conceptions of health, body, disease, recovery and medical art fail because of their own exclusiveness. The 'lesional' model predominating in objectivist medicine is unable to account for many functional diseases; it is significant that it has to qualify as 'essential' most of the etiologies of hypertension, diabetes, etc. Moreover, it runs into difficulties while trying to account for a number of especially common diseases – some infectious ones included – that remain incurable (we can think of multiple sclerosis, rheumatoid arthritis, lupus, etc.).[17] The functional model certainly manages better the cohort of 'functional' pathologies, but its exclusiveness leads it to deny the unquestionable success of scientific biomedicine, not to mention the dose of resentment that animates the patients.

Furthermore, many doctors practice both types of medicine. Even though the cohabitation proves efficient, no one clearly reflects upon it. This is the reason why, at a practical level, our conception of health and medicine becomes somewhat schitzoid.

Lastly, and paradoxically, these two conceptions of health diverge in everything but one point: the finality. As different as they may be, they both consider health to be an absolute. The contemporary world lives on Rimbaud's words: "the enjoyment of our health, spring of our faculties, selfish affection" (1972, pp. 154-155). Health has become the model of happiness. It is also the source of ethical norms. How can we explain a convergence of such diametrically opposed premises on the one hand, and such an absolutization of health on the other which goes along with the quasi-idolatrous overvaluing of medicine – or, at least, of the biosciences – in the first case and the symmetrical excessive disparagement of medicine in the other?

II. A PROPOSAL FOR A THIRD MODEL

I would like to propose the beginning of an approach to health that would solve, at least partially, those difficulties. For this purpose, I shall explore four complementary approaches.

A. *The Human Body: Principal Cause of Recovery*

The biomedical and naturalist perspectives of health are build on a double vision of the body and of bodily recovery that we could summarize as follows. According to the first view, the passivity of the sick body makes the doctor the principal or even the unique cause of health; according to the second approach, on the contrary, the patient is the essential or even the unique source of the recovery of health. Are we forced to attend the table-tennis game of this sterile dialectic, or is there a third way transcending it from inside?

The question posed here, more global than it may seem at first sight, has already been discussed by some philosophers. Indeed, Thomas Aquinas, reviving and systematizing Aristotle's categories, treated this question thematically. He did not actually treat it in itself, but indirectly, while reflecting upon the role of the teacher in the search for truth. In so doing, he used the example of medicine.[18] The text is well-known for its pedagogical content, but it is extremely rare to find it discussed for its medical content. However, I think that it is pertinent to this discussion as well. For Thomas Aquinas, the teacher is indeed (for the disciple as well as for the pedagogical art) what the doctor is for the sick person and for the medical art.[19] He poses the question: "Can a man teach and be called teacher or is this reserved to God alone?" The expression "reserved to God alone" means "reserved only to the intelligence that is God's gift within man," which excludes any help coming from outside. We can reformulate the question in a more trivial way. On the side of the receiver: does the human being need a teacher or solely the resources of his intelligence (enlightened by God) in order to learn? On the side of the giver: is the teacher really useful? This is the current debate on directivity and non-directivity in the process of learning.[20]

"There is the same sort of difference of opinion on three issues: on the bringing of forms into existence, on the acquiring of virtues, and on the acquiring of scientific knowledge" (Aquinas, *De Veritate*, q.11, a.1). Aquinas takes these three questions and sets out two opposite opinions that correspond to the two conceptions of medicine discussed above.[21] He then criticizes these two opinions.[22] Thus, we might ask ourselves: What is the truth of the matter? Aquinas states: "Therefore, in all that has been said we ought to hold a middle position between these two, according to the teaching of Aristotle"[23] (*De Veritate*, q.11, a.1). He re-introduces the tripartition and concludes that it is possible to acquire scientific knowledge.[24] But how does this happen?

Speaking about the teacher's art, Aquinas introduces the analogy with the medical art. He explains the process of the apparition of the 'new' as follows:

> We must bear in mind, nevertheless, that in natural things something can pre-exist in potency in two ways. In one, it is in an active and completed potency, as when an intrinsic principle has sufficient power to flow into perfect act. Healing is an obvious example of this, for the sick person is restored to health by the natural power within him. The other appears in a passive potency, as happens when the internal principle does not have sufficient power to bring it into act. This is clear when air becomes fire, for this cannot result from any power existing in the air. Therefore, when something pre-exists in active completed potency, the external agent acts only by helping the internal agent and providing it with the means by which it can enter into act.
>
> Thus, in healing the doctor is the servant (*minister*) of nature, which is the principal agent, by strengthening nature and prescribing medicines, which nature uses as instruments for healing. On the other hand, when something pre-exists only in passive potency, then it is the external agent which is the principal cause of the transition from potency to act.
>
> Knowledge, therefore, pre-exists in the learner potentially, not, however, in the purely passive, but in the active, sense. Otherwise, man would not be able to acquire knowledge independently. Therefore, as there are two ways of being cured, that is, either through the activity of unaided nature or by nature with the aid of medicine, so also there are two ways of acquiring knowledge. In one way, natural reason by itself

reaches knowledge of unknown things, and this way is called discovery; in the other way, when someone else aids the learner's natural reason, and this is called learning by instruction (Aquinas, *De Veritate*, q.11, a.1).

Summing up this long quotation, I want to remark that according to Aquinas, the one who takes care of the sick person is for the body what the teacher is for the soul. The conception of medicine implicitly defended by Aquinas, therefore, considers the human body as the principal cause of health and medicine as its partial cause, or more precisely, as its servant or adjuvant.[25] Aquinas draws from this view a precious therapeutic rule that can be called 'homeopathic':

> In effects produced by nature and by art, art operates in the same way and through the same means as nature. For, as nature heals one who is suffering from cold by warming him, so also does the doctor. Hence, art is said 'to imitate nature'.[26] Something similar takes place in acquiring knowledge. For the teacher leads the pupil to knowledge of things he does not know in the same way that one directs himself through the process of discovering something he does not know (Aquinas, *De Veritate*, q.11, a.1).

Applying this idea to the two classical approaches to health, one can say that both perspectives articulate only one of these two levels of causality and ignore the complex interaction between the causes. The first of the opinions Aquinas mentions corresponds to the objectivist perspective, according to which medical art is the total cause of healing. The second opinion corresponds to the naturalist perspective, according to which the human body is the principal agent thanks to its natural dynamism. Again, I am speaking here about the main representations of medicine. As stated above, physicians often implicitly integrate in their practice what they have excluded from their theories.

However, the two perspectives cannot be said to be symmetrical, since the principal causality is more of a cause than the adjuvant cause. Hence, the naturalist perspective is closer to the view proposed here, since it better acknowledges the place of the spontaneous dynamism of the human body. The objectivist perspective instead identifies the body with an insignificant, fragmented, and passive material. Moreover, complementary therapies do not hesitate to use techniques that are sometimes hard. Nevertheless, by reaction and ideology (the latter conditioning the former), complementary therapies too often deny the real but limited efficiency of biomedicine.[27] We shall see later that such a conception of the human body as the principal cause of healing presupposes a vision of nature: the one set out by Aristotle in books 2 and 3 of his *Physics*. It is necessary to reach this depth in order to understand how the three conceptions of health oppose each other.

This first approach to health is not yet sufficient for a proper account of human health. It stresses the rooting of health in a dynamism that precedes it and accompanies the science and the art of medicine, but it does not actually give a definition of health. A further dimension needs to be considered.

B. Health as Unified Body

Health is a characteristic of the human body. I mean here that being healthy characterizes the organism in itself and in its distinction to the soul. On the one hand, the philosophy of nature establishes the dual structure of living beings in general and of the human being in particular, avoiding the double error of monism and dualism.[28] On the other hand, experience shows that one may be physically unhealthy while one's psyche remains intact. Psychophysics has rightly emphasized that any pathology and even any inner tension causes repercussions or is expressed physically; but the assertion cannot be reversed: not every organic disease always has its counterpart in a psychical dysfunction.[29]

As I have argued elsewhere (Ide, 1996, pp. 143-156), the human body is at the same time one and open: it is one precisely so that it may be open. Moreover, health is the property of a unified body. This proposition can be validated at the level of the body-subject as well as at the level of the body-object. At the level of subjective experiences, the healthy body is experienced as silent (Leriche, 1936) or as the bearer of well-being (WHO, 1946). Peace or harmony are indeed the fruits of unity (cf. Ide, 1998, pp. 67-68). This is also true at the objective level. As a matter of fact, the human organism is the most complex and the most harmonious of all living bodies. This diversity needs to be unified: the multiplicity of organs and functions must be adapted to the common good of the organism. This is the proper characteristic of health: it is unifying, that is to say, synthetic, as the doctor and philosopher Hubert Saget states: "while normality is synthetic, pathology is analytical. It isolates, dissociates, and appears first as a factor of rupture of integration. The achievement of an organ is to be forgotten as a distinct element and to contribute in such an 'integrated' and 'transparent' way to the cohesion of the whole so that only such a whole seems to exist" (1976, p.123). Negatively, a body can be healthy without being open, for instance, without giving life or welcoming others.

Thus, one might conclude that health can be defined as the property, or more precisely, as the quality[30] of an harmonious and unified body made for openness. The Aristotelian concept of quality-disposition coincides with the notion of capacity that some people use to define health: "health is the capacity that every human being possesses to face his environment or to assume responsibility for its transformation" (Illich, in: Dufresne, 1985, p. 989). "The healthy man is the one who is able to respond to a complex and changing world, who is able to invent at every instant new forms of behavior, who is able to modify the world rather than to adjust to it." (Bastide, in: Dufresne, 1985, p. 989). "By man's health, we understand the *capacity to face all situations that would aim in themselves at weakening his potential and his dynamism of life, while maintaining the highest possible degree of autonomy and of capacity to act*"(Verspieren, 1984, p. 368, italics original).

This is the reason why the approach to the normal (and the pathologic) given by Canguilhem (1966), although pertinent, does not seem sufficient to me. If one speaks too much about adaptation and openness to the environment, one ends up forgetting that life, health, and the body possess a proper unity. Biomedicine legitimately objectivizes this unity, for which the consciousness of the body-subject accounts; the body-subject never intermingles with any of the other surrounding bodies. More generally, the concept of adaptation became central in biology after Darwin, because the relational distinction living being/environment was privileged to the prejudice of the hierarchical distinction between inert, vegetative, animal and human beings. These two distinctions, instead of excluding each other, are actually complementary. The analysis of the third dimension of health will confirm this.

This approach to health also remains insufficient, for three reasons. First, it does not directly take into account what has been said about the natural dynamism of the human body, nor does it shed light on the place of technique in the healing process. Moreover, it does not explain the reasons for the current absolutization of health. Finally, it does not explain the two models of health, objectivist and naturalist. This is a result of its lack of precision with regard to the kind of openness that is at stake. Indeed, the openness can be twofold, according to the perspective one adopts: it may be considered from above (as the original gift) or from below (as the self-giving), as the next section will show.

C. Health as Appropriated Body

Elsewhere I have unfolded an anthropology and even an ontology of the gift.[31] I have argued that the act of giving is a dynamic process that includes three stages: the reception of the gift (the gift *for* the self); the appropriation of the gift (the gift *to* the self); the offering of the gift (the gift *of* the self). In more concrete terms, any finite reality (or, from a monotheistic perspective, any *created* reality) follows a trinary rhythm: reception (or donation, according to the standpoint one chooses), appropriation, and diffusion.

Let us apply this trinary rhythm to the human body. First, the body is *received*, which means that it bears inborn dynamical processes that it did not give to itself. This is, for instance, the case with the intracellular processes of translation and transcription, or with the stimulation of specific cones by a given luminous radiation. In negative terms (according to a certain view of material nature), these biological laws cannot be modified in themselves.[32] In positive terms, they are a received and inherited dynamism that aims at the service of the body considered in its wholeness.

Second, precisely because it is inherited, this dynamism must be welcomed, *appropriated* by the consciousness and the freedom of the person. Such an appropriation corresponds to the unity mentioned before and presents the two aspects, subjective and objective.

Third, the human body is not subordinated to itself: it is made to be *offered*. More precisely, it is an *organon* (to use the expression *instrument* would be too dualistic) subordinated to the superior good of the soul,[33] which is itself made for openness to the other and, ultimately, for friendship and communion (cf. Haegel and Rouvillois, 1995).

If one considers the case of living bodies, and especially of the human body, health seems to qualify the body in its intermediary stage, namely the appropriation. A healthy body is an appropriated body. I see three main reasons for that. To begin with the most evident: it is one thing to be healthy and another to give oneself. We know very well that we can live by caring about our physical form only, without being even minimally open to others. We can even appeal to our health as a pretext to refuse generosity. This is the case for some women who refuse to have a child (to *give* life) in order to preserve their figures and avoid stretch marks. A 'given' body is a tired body that does not fear the wrinkles of wear. More deeply, we face here two diverse intentionalities: the appropriation of the body directs the body to the self; the gift directs it to the other.

In another respect, "a healthy man is not a man whose health would have been fabricated," Gadamer says (1998, p.44). The human body is inhabited by some physiological processes, a harmony that man did not create, although he becomes responsible for it. A corporeal dynamism, which can be said to be natural insofar as it is given, precedes freedom. The body is anterior to any objectivization and even to any appropriation. The body-nature comes before the distinction between body-object and body-subject.

On the contrary, it is one thing to receive your body and another to appropriate it to yourself. Is it sufficient to be inhabited by some biological dynamical processes to be healthy? We know that for human beings the right use of the body is not ruled by instincts: human beings must choose their food, their rest, etc. They need to become conscious of their bodies, take care of them in a responsible way and integrate them within a project.

Lastly, and more profoundly, whether we refer to health-subject or health-object, the healthy human being calls for consistency, for what Blondel would have called stability (cf. Tourpe, 2000, pp. 14 ff.) In the former case, this stability is subject-like (at home): to be healthy is to feel well at home. In the latter case, this stability is defined as substance (*in se*): to be healthy is to conform to a certain number of quantifiable criteria, since the corporeal substance is material and therefore measurable. As the consistency of reality changes, so too does its donation change (in the sense of foundation or diffusion); likewise, substance to operation or relation, unity to openness, and what is present to its presence. Health emphasizes how important it is to stabilize the presence in the present, the donation in the effective gift. Freedom does not convert the original gifts of nature that are the body and its inborn dynamical processes into a generous self-giving if there is no prior appropriation of that body as mine. This is my definition of health. After being son and before being father, one must be friend of oneself. Any self-giving presupposes a self.

Such a definition of health as appropriation of the body takes into account the preceding definition, which identified health with the unified body. In fact, the appropriation is a work of integration, and to speak about integration is to speak about unity. It also does justice to the spontaneous dynamism, the activity of the body corresponding to the first moment of the gift. Indeed, the body did not give this active capacity to itself; it received it. Finally, it situates the contribution of medical technique, which intervenes in the second moment of the gift, and then presupposes the presence of the organism's dynamism. Even though it is formulated in a different way, we can find here also the hierarchical distinction between the principal and the adjuvant causes in the process of recovering health, namely the human body and the medical art respectively.

Nevertheless, some could object that these distinctions are irrelevant. Does experience not show that the more we give ourselves, the more we preserve ourselves? Someone who is too anxious about his health and about himself is often a hypochondriac, who will soon become sick with imaginary pathologies. Such an objection rightly stresses that the three moments of the body-of-gift (here the moments 2 and 3 are closely and vitally united). However, I would respond that interconnection does not mean identity. Health is a condition for self-giving, as unity is the foundation of openness. Nevertheless, the condition is not what is conditioned, nor is the foundation what is founded.

As has already been mentioned, this new approach to health sheds light on the value and the deficiencies of the two previous conceptions of health. The objectivist approach neglects the first moment of the gift. When considered in its concrete integrality, the body is not only extended but also active and dynamic. More generally, the original gift is the gift inherited by human freedom. It cannot be constructed but is the condition and the material for any further construction. It has been said that man does not exert any power over his physiology, which is part of the involuntariness that Paul Ricoeur (1950, p. 456) has shown is compounded with voluntariness.[34] However, the different moments of the gift are bound to each other. Hence, if one neglects the receptivity of the corporeal dynamism, one falsifies the moment of appropriation of the body. To formulate it more precisely, the activity of consciousness and freedom, which are the agents of the appropriation, is valorized proportionally to the inertia and the presumed passivity of the corporeal material. Consciousness imposes its laws (its law) on this material; freedom hails and controls it, and soon desire (mixed with freedom) projects its omnipotence onto it. The spirit can realize the relation of domination it dreams about thanks to medical technique. This is the reason why the objectivist model overvalues technique and an extreme specialization *versus* the spontaneous and unified dynamism of the body.

On the other hand, the naturalist approach to health does justice to the first moment of the gift. Nevertheless, the Hegelian dialectic of mastership and slavery warns us about the necessary reversion of the relation of domination. Unjustly enslaved by the spirit thanks to the mediation of biomedicine, the body rebels and takes revenge with naturalist medicines so that any legitimate authority of its former master is refused. The classical medical practice is then criticized in the same harsh and unilateral way.

One of the above posed questions remains unsolved: *Quid* of the current absolutization of health and of the strange complicity between the two dominant conceptions on this point?

D. A Cosmo-anthropo-theology of Health

Some will be puzzled or even disturbed by this title.[35] However, it seems to me that if we want to understand what is ultimately at stake with health, we must appeal to a properly religious vision or rather to a reinsertion of health within a complete vision of reality that would integrate the three regional ontologies: God, man, and nature.

1. The contemporary idolatry of health
I shall begin by considering a first aporia. Although it is inscribed within the finitude of the human body, health now has an infinite amplitude. This infinite (this bad infinite, Hegel would have said) is twofold. At the level of subjective investment, our society is inhabited by an "inextinguishable thirst" for perfect health, as Illich noticed (1975, p. 206). Every unique, obsessional, and compulsive investment saturates the field of consciousness that is open to the infinite. Nevertheless, the contemporary preoccupation with health is hypochondriac. It is significant that the main wish for Christmas and the New Year is "good health."

However, health in its objective content has been idolized. It has become a substitute for God from two standpoints: the term and the origin. We look today for a health without deficiencies. The object of health is the perfect man. We aspire to a body without suffering, lesion or pain. Imperfections in physical appearance, menopause, or growing old are considered diseases. The logical consequence of this is the growing intolerance of patients with regard to the smallest failing of the medical staff. Disappointment (and soon mistrust[36]) are equal to the expectation. Continuing his stimulating critique of contemporary utopias and ideologies, Lucien Sfez analyses the new paradigm of perfect health "as the sole and unique world project." "The Great Health, as an ideological, symbolical, and utopian construction [...] has for itself all the beginning of the third millennium" (Sfez, 1995, p. 24).

Why should we criticize such a project? Sfez analyzes the visions underlying three worldwide enterprises. The first is the sequencing of the human genome, about which I shall speak later. The second is the American operation *Biosphere*, which isolates under large hangars eight human beings and three thousand four hundred animal and vegetable species for two years. The third is the project *Artificial Life* of the Santa Fe Institute, which intends to create artificial life in a computational way. This last project aims at constructing a "bio-eco-religion", thanks to the possibilities granted by technoscience.

In the project *Biosphere II*, scientists dream about a perfect Earth that will sooner or later supplant the imperfection of our first Earth where unconscious human beings pollute everything. "I had the opportunity to speak with 'Biospherians' and among them with a British lady, who was feverishly telling me that she had lived an extraordinary experience: the relationship with her body changed; she breathed pure air and ate the food she was producing with her own hands, etc. She was explaining to me that her companions and herself had 'a superior and perfect health' because they had an extremely low cholesterol level" (Sfez, 1997, p. 46). If we used to think that the imperfections of man came from the environment (infection, pollution, stress, etc.), we think today that they are more profoundly rooted in his genome. Indeed, we now know that we all bear approximately three hundred defective genes that are not expressed but which, once they are no more repressed, can be the source of diverse serious pathologies like cancer, multiple sclerosis, etc. It is necessary, then, to modify man's chromosomal material, that is to say his material *origin*. Man will be perfect only when we are able to master this origin. Hence the passion, mixed with fear, for the project of sequencing the human genome. Today, the biotechnologist dreams of taking up God's project, who created man after His image. However, strengthened by the knowledge he acquired in genetics, he does not want to potter any more, he wants to eliminate *bad genes* or even genes responsible for the fashioned notion of cellular suicide (cf. Ameisen, 1999).

We move from a man created after the image of God to a man created after the image of man and arrogating the divine prerogatives to himself. Does this not recall the project of Doctor Frankenstein? In an important work, Howard Segal (1985) describes the utopian American technology in the years 1880-1930. The author enumerates no fewer than 160 technological utopias during this period. Those who created these utopias were not philosophers or men of letters, but practical men (managers, doctors, and engineers) whose goal was to improve immediately the life of their fellow-citizens. Some utopias have been quite successful and sold millions of specimens. This was, for instance, the case with Dr. Kellogg. Heard about him? He was the inventor of *Cornflakes*. We often ignore the fact that this doctor wanted to produce aliments to help the American youth, and then the youth of the whole world, to become robust and healthy.

The utopian dream is clearly one of a new humanity. Current biomedicine seems to give to this dream the means to acquire its content. The project of fabrication of an adamantine being endowed with the divine privilege of immortality joins the double divine property of mastership of the origin and of perfection of the goal.[37]

For these three reasons, saturating, perfect, and fully mastered health occupies in our imaginings and intentionality the place usually reserved for God. *Our current search for health*, be it objectivist or naturalist, *is idolatrous*, in the rigorous, biblical and theological sense of the term: the idol is a divine substitute. Health becomes a creature believed to be the Creator. "In order to speak about health in 1999 the search for health has to be understood as the reverse of salvation, it has to be understood as a societal liturgy serving an idol that extinguishes the subject." (Illich, 1999).

Pope Pius XII, stressing the value of the more extended, deep and positive definition of health given by the World Health Organization in June 1946, said: "in your opinion health positively comprises the spiritual and social well-being of humanity and, in this respect, is one of the conditions of universal peace and common security" (Pius XII, 1949, p. 98). It is significant that the Pope did not quote the whole expression: "state of complete well-being" (WHO, 1946). Later, he added that health "is part of the conditions for the dignity and the total good of humanity, for its corporeal and spiritual, temporal and eternal good" (Pius XII, 1946, p. 99). Note that "condition" does not mean "constituent".

Is a purely secularized and methodologically atheistic approach to health sufficient to account for this complex reality? Is it enough to recall that Canguilhem once said that "a perfect and continuous health is an abnormal fact" since "disease is actually included in the experience of living beings" (1966, p. 86)? Will this be sufficient to exorcise this compulsive and idolatrous desire for perfect health? The following fact will amplify these questions.

2. The double form of the idolatry
As has been emphasized above, both the biomedical and the naturalist conceptions of health strangely oppose each other in numerous aspects except one: the 'infinitization' that we have just been discussing. How shall we explain this convergence?

In his *Carnets*, Wittgenstein makes this surprising statement: "there are two divinities: the World and my independent I" (Wittgenstein, 1971, p. 142). Similarly, the idolatry of health manifests itself in a twofold way. In the objectivist model, infinity is attributed to health *via* technique or technical reason. Reason's power to command a nature conceived as passive and insignificant is infinite. Sartre (1947) considered that Descartes had attributed to human freedom the attributes of divine freedom. We have inherited from Descartes the famous project to become "like masters and possessors of nature" (1950, p. 168). Given that the human body constitutes for human beings what is natural, technical improvements made the modesty of the expression 'like' disappear.

Within the naturalist model, the 'infinitization' moved to the pole of the natural. It is nature, embodied in man, that becomes the bearer of divine promises. It is well-known that 'New Age' and 'Deep Ecology' developed a pantheistic approach to nature; the project Gaïa gives to Earth the name of a Greek divinity (Lovelock, 1979). In his book *De l'homme-cancer à l'homme-Dieu*, Bernard Woestelandt (1986), a medical doctor, proposes to go from man-Cancer, who is pure ignorance, to man-God, who is the man of full consciousness, and hence, of divinity. With a presumptuousness which may provoke either stupefaction or misgiving, the back cover of the book dares to say: "At the end of the 20th Century and at the dawn of the age of Aquarius, all that was hidden is revealed. Recent discoveries of modern physics bring something new to the interpretation of Sacred Texts and a synthesis is outlined by vanguard theologians, physicists and physicians." An equation sums up the whole book: "God = Love = Freedom = Knowledge = Wisdom" (Woestelandt, 1986, p. 364).

The biomedical conception, then, deifies the spirit (reason and freedom) of the human being, while the naturalist vision deifies the body (as the holder of the wisdom of nature). Each of these two models takes to the infinite degree one of the two aspects of reality between which knowledge has been developed and then torn for the last four centuries, according to the famous diagnostic of Edmund Husserl (1976, 1989).

Is a reconciliation possible? Hegel taught us that we need a third pole, apart from nature and the finite spirit: the infinite Spirit, that is to say God, who is the only absolute. Putting into a philosophical perspective the history of the immediate Hegelian posterity sheds light on our topic.[38]

What happens in the case of atheism? Only man can become the center. Nevertheless, we cannot avoid all that we know about God. More precisely, all the attributes of the Other will be found in man, replaced in the sphere of man's essence. This is what happened to one of Hegel's students, Feurbach (who should be seen as a student rather than a disciple of Hegel). His point of departure was the finitude of man. He asserts that man "can never overcome his true essence (*Wesen*)" (1982, p. 128). As a matter of fact, all that man can conceive about other beings, for instance extraterrestrial minds (in Feuerbach's example), always comes from his point of view: "these are always determinations coming from his own essence" (Feuerbach, p. 128).[39] What is true for any object, particularly applies to the religious object. If the sensible object is external to man, the religious object is internal. Hence, *"the awareness of God is the self-awareness of man, the knowledge of God is the self-knowledge of man"* (Feuerbach, 1982, p. 129-130; italics added). From now on, man will recover for himself all that was said about God, on behalf of the Man he has to give rise to.

However, man cannot face himself. He needs to have someone opposite to him. As a matter of fact, the divine alterity dissolves in the sameness of the human identity. Any becoming of the human essence becomes mystification, pure appearance: it is just coming back to the self. It is necessary to leave the human sphere. Since the one opposite to man cannot be either the same as man nor the completely Other who was banished, only *nature* remains. Hence, the new dialectic becomes that of man and nature, of humanism and naturalism. However, it is still necessary to find an effective mediation, in order not to remain in instability and, sooner or later, to fall into one of the extremes. Feuerbach understood that, but he could not resist the swing. His humanism quickly inverted to a 'vulgar' naturalism: "Man is what he eats!" Still too busy with assuming the religious heritage, and then too reactive, Feuerbach was not able to adopt a positive position towards nature. Later, Marx would think of reconciling man with nature by suppressing the question of God.[40]

III. CONCLUDING REMARKS

The history of thought shows that we cannot neglect one of the three poles (God-man-nature) without falling into a mortal face-to-face that absolutizes one of these poles. Moreover, the three poles do not have the same place: God, infinite being, faces the double finitude of nature and of the finite spirit that is man.

What is true for nature in general applies also to the human body and to physical health. Nature (understood as the negation of spirit, to speak in Hegelian terms) is for a finite mind what the body is for the soul in the human being. Nowadays, the question of health is posed in a secularized context. This is the reason why the conceptions of health reproduce, at their own level, the more general dialectic man-nature. We then realize that the only way to tear ourselves away from the temptation to absolutize health does not consist in "auto-limiting" it, in preaching to it a kind of stoic restraint, which, by stressing one of the two poles, technical reason or nature, would only strengthen the other one. Instead, it consists in giving anew to God the infinity that belongs to him alone, and to the Caesar of medical power and of naturalist medicine its proper place, namely finitude. A balanced vision of health cannot risk understating an integral understanding of reality articulating God, man, and nature.

There is an objection that will allow me to establish a correlation between the approach to health I am proposing here and the preceding ones. Why should a philosophically adapted approach to health refer to God? Is the trinary logic of the gift not sufficient? As a matter of fact, to ground medical practice on an adequate vision of the original dynamism of the body allows us to avoid the first form of absolutization of health, which is proper to the idolatry of technical reason present in the objectivist conception. On the other hand, acknowledging the pertinence of the biomedical approach allows us to avoid the second form of absolutization of health, which is proper to the idolatry of the nature-body present in the naturalist conception. In other words, we have to interlace the original gift with the gift to the self. A fact that is valid for the two models is that health aims at self-giving. And to have an aim means to be finite. Hence, health definitely has to escape the temptation of absolutization.

A precise answer to the question stated above would require a long analysis. Thus, I shall conclude here by proposing just a reflection. I think that the openness of the healthy body to an Other who is more than the self definitely liquidates the utopia of the infinitely healthy body. However, on behalf of what could we prescribe this third moment of the gift? It is significant that the objectivist and naturalist models consider self-giving as something accidental to the body and to health. In fact, as I have argued elsewhere, human beings are led to self-giving by the logic of superabundance, and not merely by one of debt or counter-gift (Ide, 1997, especially ch. 24). Aware of the fact that bodily health is a gift, the human person freely enters into the dynamic of the gift, surrendering the body to the dynamic of love and life. The primacy of technical power offends the original character of the donation.[41] Even the grounding of the body in nature is not sufficient to recognize that a gift is at stake. One should still discern behind the immanent gift of natural dynamic processes the free generosity of a transcendent Giver.

Official Vatican Congregation for Catholic Education
Rome, Italy

NOTES

[1] Translated from the French by Stéphane Agullo, in collaboration with Paulina Taboada. All quotations have been translated from the original text in French, except the one from Aquinas, which corresponds to the official English translation of *De Veritate*.

[2] What I refer to here as the 'naturalist' model of health is also known as the 'holistic', 'organismic', or 'systems' conception in the current debate on the concept of health.

[3] However, French possesses one neglected linguistic resource, *pathologie*, which refers to the objective dimension of disease.

[4] According to Benoist, *sickness* signifies "the process of socialization of *disease* and *illness*" (1983, p. 51). This conception sheds light on the debate between Eisenberg (1977, pp. 9-23) who places *illness* on the side of the disease-subject, and Fabrega (1977, pp. 201-228; 1978, pp.11-24) who situates it on the side of the disease-society. It also clarifies the position of Genest (1981, pp. 5-19).

[5] For this point as well as for the following, I refer the reader to my essay: *Le corps à cœur. Essai sur le corps* (Ide, 1996), especially part I., chap. 3 – 6, where I show that one can qualify this model of the body as *mechanist*.

[6] Dreaming about the evolution of medicine in this century, Jean Bernard (1973) imagines a doctor who falls asleep in 1960 and wakes up in 1990. The description Bernard makes is visionary. "How to foresee and organize the necessary adaptations? [...] Modern man, dissected in organs by the anatomist, cut into tissues and cells by the histologist, pulverized in molecules by the biochemist, volatilized in electrons, protons and neutrons by the physicist, presents himself under the form of a cloud of elementary particles. Behind this fragmentation, the doctor finds again or maintains the always renewed but constant unity of his patient" (Bernard, 1973).

[7] According to Dagognet, organ transplantation sees the body as "a simple assembly of pieces, which can themselves be changed in a standardized way" (1998, p. 282).

[8] "Any portion of our body or any link of our physiological internal chains would depend on one locus, localized on one of our chromosomes, which would offer us an image cut into tiny pieces, like a mosaic (a division into a thousand pieces)" (Dagognet, 1998, p. 284).

[9] A detailed story of 'naturopathy' can be found in Carton's (1943).

[10] For a comment on some stages of the transition from the objectivist to the holistic-naturalist conceptions of health, see my postface to Theillier and Geffard (2001, pp. 341-353).

[11] This is the way in which the American anthropologist Byron Good (1998, p. 74) opposes "the occidental medicine [that] considers the body as a complex biological machine" to the Zinacantanal medicine, in the region of Chiapas in Mexico, for which the disease is "one aspect of the person considered as a whole and in its relationship to the society and the supernatural." Good builds his analyses upon the work of the two ethnologists Fabrega and Silver (1973).

[12] It is interesting to note that François Laplantine (1986) approaches health only at the end of the healing process and not in itself.

[13] In the second part of Laplantine's work (1986, pp. 53-175) one realizes that his four approaches to disease roughly follow Aristotle's four causes. In fact, the first distinction is taken from the subject's side (material cause), the second from the origin (efficient cause), the third from the nature of the disease (formal cause) and the fourth from the sense (final cause). It is precisely in this way that Laplantine's work shows a true broadness of mind.

[14] One could have demonstrated this point starting from the four elementary forms of disease described in the third part of Laplantine's work (1986, pp. 177-232).

[15] See, for example, the fundamental work by Keit Thomas (1971).

[16] What is true for Babylon applies even more to Egypt (cf. Bardinet, 1995).

[17] This aspect has been developed in the work of Theiller and Geffard (2001).

[18] The expression 'example' is used here in the Aristotelian sense of analogy.

[19] There is more than a simple analogy here. In the modern West, the objectivization of the sick body has been correlated to the training of the body through education for a long time (cf. Foucault, 1975a).

[20] I am referring here to the question *De Magistro* from the *Q. D. De Veritate*, q.11, a.1. This text has the advantage of enlightening medicine and, correlatively, health from the highest metaphysical principles. Cf. also *Scriptum super Sententiis*, Book II, d. 19, q. 1, a. 2, ad 4um; *ibid.*, d. 28, q. 1, a. 5, ad 3um; *Summa theologica*, Ia, q. 117, a. 1; *Summa contra Gentiles*, II, chap. 75.

[21]   In spite of its length, and because of its richness, Aquinas' original text (*De Veritate*, q.11, a.1) deserves to be transcribed here. The quote is taken from the essential parts of the *respondeo*, intercalating some subtitles:

First opinion:

Some have said that all sensible forms come from an external agent, a separated substance or form, which they call the giver of forms or agent intelligence, and that all that lower natural agents do is prepare the matter to receive the form. Similarly, Avicenna says in his *Metaphysics* [IX, 2] that "our activity is not the cause of a good habit, but only keeps out its opposite and prepares us for the habit so that it may come from the substance which perfects the souls of men. This is the agent intelligence or some similar substance." They also hold that knowledge is caused in us only by an agent free of matter. For this reason Avicenna [*Liber De anima Seu Sextus De Naturalibus*, V, VI] holds that the intelligible forms flow into our mind from the agent intelligence.

Second opinion:

Some have held the opposite opinion, namely, that all three of those are embodied in things and have no external cause, but are only brought to light by external activity. For some [Anaxagoras, whose position is reported by Aristotle, *Physics*, I, 4, 187 a 26] have held that all natural forms are in act, lying hidden in matter, and that a natural agent does nothing but draw them from concealment out into the open. In like manner, some [John Damascene, *On Orthodox Faith*, III, 14] hold that all the habits of the virtues are implanted in us by nature. And the practice of their actions removes the obstructions which, as it were, hid these habits, just as rust is removed by filing so that the brightness of the iron is brought to light. Similarly, some [namely those who defend the theory of reminiscence that Thomas, following St Augustine, attributes to Plato. Cf. St Augustine, *De Trinitate*, XII, 15, 24] also have said that the knowledge of all things is con-created with the soul and that through teaching and the external helps of this type of knowledge all that happens is that the soul is prompted to recall or consider those things which it knew previously. Hence, they say that learning is nothing but remembering.

[22]   *Criticism of the two opinions:*

But both of these positions lack a reasonable basis. For the first opinion excludes proximate causes, attributing solely to first causes all effects which happen in lower natures. In this it derogates from the order of the universe, which is made up of the order and connection of causes, since the first cause, by the pre-eminence of its goodness, gives other beings not only their existence, but also their existence as causes. The second position, too, falls into practically the same difficulty. For, since a thing which removes an obstruction is a mover only accidentally [Aristotle, *Physics*, VIII, 4, 254 b 7 and 255 b 24], if lower agents do nothing but bring things from concealment into the open, taking away the obstructions which concealed the forms and habits of the virtues and the sciences, it follows that all lower agents act only accidentally (Aquinas, *De Veritate*, q.11, a.1).

[23]   Aquinas refers here to Aristotle, *Physics*, I, 8, 191 b 27 and III, 1, 201 a 9 and 201 b 5; see also *Metaphysics*, XII, 5, 1071 a 5.

[24]   *What is the truth?*

For natural forms pre-exist in matter not actually, as some [the defenders of the second opinion] have said, but only in potency. They are brought to actuality from this state of potency through a proximate external agent, and not through the first agent alone, as one of the opinions [the first one] maintains.

Similarly, according to this opinion of Aristotle [*Nichomachean Ethics*, VI, 13, 1144 b 4 and II, 1, 1103 a 24], before the habits of virtue are completely formed, they exist in us in certain natural inclinations, which are the beginnings of the virtues. But afterwards, through practice in their actions, they are brought to their proper completion.

We must give a similar explanation of the acquisition of knowledge. For certain seeds of knowledge pre-exist in us, namely, the first concepts of understanding, which by the light of the agent intellect are immediately known through the species abstracted from sensible things. These are either complex, as axioms, or simple, as the notions of being, of the one, and so on, which the understanding grasps immediately. In these general principles, however, all the consequences are included as in certain seminal principles. When, therefore, the mind is led from these general notions to actual knowledge of the particular things, which it knew previously in general and, as it were, potentially, then one is said to acquire knowledge (Aquinas, *De Veritate*, q.11, a.1).

25 Dominique Foldscheid comes back to this idea when defining medicine as "operation of mediation, or, playing with the words, of re-mediation. ...the activity of re-mediation has nature as a field, but nature understood in its dimension of *telos*, of internal auto-finality" (1996 pp. 506-507). It is interesting to note that his understanding of *telos* coincides exactly with the Aristotelian meaning of it. Precisely because of that he also distinguishes current technoscience from the ancient *technè*: "Considered in its essence, medicine is neither a science nor a technique, but a healing practice accompanied with science and instrumented with technical means" (1996, p. 509).

26 Aquinas refers here to Aristotle, *Physics*, II, 2, 194 a 21.

27 According to the logic of *ressentiment* analyzed by Nietzsche and re-read by Gilles Deleuze (1962).

28 I have developed this idea elsewhere (cf. Ide, 1996), especially the Part II, chap. 5.

29 On this topic, see the small, fortifying book by Anselm Grün and Meinrad Dufner (2000). The authors clearly distinguish physical disease as an expression of the soul (pp. 12-27) and as a chance, i.e. as a spiritual occasion, without psychical cause (pp. 27-40).

30 Aristotle included health (more precisely the health of the body) in the first species of the category 'quality', namely, among those states and dispositions that "can be easily moved and quickly changed, as heat and cooling down, disease and health" (*Categories*, 8, 8 b 35-37).

31 For a more pastoral perspective cf. Ide, 1997; for an anthropology and ethics of the 'gift' cf. Ide, 1998, p. 39 – 58; and for the application of this concept to the body, cf. Ide, 1996, part II., chap. 7.

32 What I mean here is that even though human freedom may in some exceptional cases and by ascetic means induce a decrease in the pulse or breathing rates, as for instance in the case of yogis, this does not mean that the brain could work without oxygen; in other words, human freedom cannot change the laws of nature.

33 Aristotle said that the soul is not only the efficient origin and the formal substance of the body, but also its aim. (*On the soul*, II, 4, 415 b 10-11)

34 Paul Ricoeur (1950, p. 456) speaks about "the condition of a will reciprocal of an involuntariness," and about a choice "paradox of initiative and receptivity" (1950, p. 456 and 454).

35 I recall here the 'cosmotheandrism' or the 'theanthropocosmism' about which Raimon Panikkar has spoken: "As a matter of fact, God, Man, and Matter connote three real aspects that are not autonomous, but intimately connected and that are always to be grasped together

in their very distinction. They are constitutive dimensions of reality. [...] Between these three dimensions, there is then what we can call a relation of *autonomy* – but an autonomy whose content is an *inter-in-dependence*" (1998, p. 17). I am adopting here the aim of Panikkar's argument, but not necessarily the detail of his development of it.

[36]   This mistrust is also linked to the objectifying logic proper to capitalism which made health become merchandise, a product of the free-market economy. "There is a progressive passage towards a free-market economy's medicine which then is doomed to juridical or even judiciary tilts. Trust towards the doctor is replaced by mistrust, the same mistrust a consumer experiences in front of a product" (Sicard, 1993, p. 118). On this point see also Boltansky and Chiapello, 1999 (especially pp. 501-576).

[37]   Michel Houellebeck (1998) shows this in a sarcastic and pessimistic way in his successful novel, which – in spite of its obscenity – can be considered as visionary.

[38]   On this topic see Claude Bruaire's (1985) interesting text. I am also referring here to André Leonard's interpretation of the three Hegelian syllogisms at the end of *The Encyclopedia of Philosophical Sciences*, §575-577 (in Leonard, 1980, pp. 40–48).

[39]   One realizes that this is a noetic reduction and that this noesis is idealistic. "*Without object, man is nothing*" (Feuerbach, 1982, p. 120; italics added by the author). "You know man from the object; in it, his essence *appears to you*" (ibid., p. 121).

[40]   Hegel, in fact, goes further while saying something a little bit different: according to him, reconciliation is not operated by God as such but by Christ. Indeed *The Phenomenology of Mind* explains that man is just reduced to being an unhappy consciousness in front of God. Only Christ reconciles finite spirit and infinite spirit: his mediation exorcises any unhappy consciousness. Then, the consequences described above do not come from atheism as such but from the collapsing of Christianity: when the Incarnate Word disappears, the relationship man-God comes back to its pre-Christian fragility. Thought is then shaken between a theism indifferent to man and a humanism proclaiming its self-sufficiency. We see therefore the determinedly atheistic humanism that appears.

[41]   See Alain Finkielkraut's (1996), which refers to the first edition (1951) of what later became *Les origines du totalitarisme* by Hannah Arendt (cf. Clavier, 2000, p. 291-292.).

BIBLIOGRAPHY

Augustine: 1970, *De Trinitate*, The Catholic University of America Press, Washington.
Ameisen, J-C.: 1999, *La sculpture du vivant*, Seuil, Paris.
Aquinas, T.:
—— 1994, *De Veritate*, translated by James V McGlynn, S. J., Volume II, Hackett Publishing Company, Indianapolis/Cambridge.
—— 1947, *Scriptum super Sententiis*, Lethielleux, Paris.
—— 1961, *Summa contra Gentiles*, Marietti, Turin.
—— 1894, *Summa theologiae*, Forzani and Sodalis, Rome.
Aristotle:
—— 1973, *Categories*, Harold P. Cooke (trans.), The Loeb Classical Library, William Heinemann LTD and Harvard University Press, London/Cambridge (Massachusetts)

—— 1969, *Metaphysics*, Books X-XIV, Heigh Tredennick (trans.), The Loeb Classical Library, Harvard University Press and William Heinemann LTD, Cambridge (Massachusetts)/London.

—— 1975, *Nicomachean Ethics*, H. Rackham (trans.), The Loeb Classical Library, Harvard University Press and William Heinemann LTD, Cambridge (Massachusetts)/London.

—— 1980, *Physics*, Volumes 1 and 2, Philip H. Wicksteed and Francis M. Cornford (trans.), The Loeb Classical Library, Harvard University Press and William Heinemann LTD, Cambridge (Massachusetts)/London.

Avicenna:

—— 1972, *Liber De Anima Seu Sextus De Naturalibus*, Ed. S. van Riet, Leiden.

Bardinet, T.:

—— 1995, *Les papyrus médicaux de l'Egypte pharaonique*, Penser la médecine, Fayard, Paris.

Basil the Great: 1857, *The Longer Monastic Rules*, in *Patrologia Graeca*, Migne, Paris.

Benoist, J.: 1983, 'Quelques repères sur l'évolution récente de l'anthropologie de la maladie', *Bulletin du séminaire d'ethnomédecine*, 9, 51-58.

Bernard, J.: 1973, *Grandeur et tentations de la médecine*, Buchet-Chastel, Paris.

Bottero, J.: 1984, 'La magie et la médecine règnent à Babylone', in *Les maladies ont une histoire*. L'Histoire, Seuil, Paris.

Boltansky, L. and Chiapello, E.: 1999, *Le nouvel esprit du capitalisme*, Gallimard, Paris.

Bruaire, C.: 1985, *La dialectique*, Que sais-je ? n° 363, PUF, Paris.

Canguilhem, G.:

—— 1966, *Le normal et le pathologique*, Quadrige, PUF, Paris.

—— 1992, *La connaissance de la vie*, Bibliothèque des textes philosophiques, Vrin, Paris.

Canto-Sperber, M.: 1998, 'Anne Fagot-Largeault: l'éthique et la médecine', *Magazine littéraire*, 361, 102.

Carton, P.: 1943, *L'apprentissage de la santé, histoire d'une création et d'une défense doctrinale*, Maloine, Paris.

Clavier, P.: 2000, *Le concept de monde*, Philosopher, PUF, Paris.

Dagognet, F.: 1998, *Savoir et pouvoir en médecine*. Les empêcheurs de penser en rond. Institut Synthélabo, Le Plessis Robinson.

Damascus, J.: 1898, *Exposition of the Orthodox Faith*, Post Nicene Fathers, Volume IX, Series II, Aberdeen.

Deleuze, G,: 1962, *Nietzsche et la philosophie*. PUF, Paris.

Descartes, R.: 1950, *Discours de la méthode*. In: Bridoux, A.: *Œuvres et lettres*, Bibliothèque de la Pléiade, Gallimard, Paris.

Dufresne, J.: 1985, 'La santé'. In: Dufresne, J., Dumont, F. and Martin,Y.: *Traité d'anthropologie médicale. L'institution de la santé et de la maladie*, Presses de l'Université du Québec et Institut québécois de recherche sur la culture, Québec / Presses Universitaires de Lyon, Lyon, pp. 985-1013.

Eisenberg, L.: 1977, 'Disease and illness,' *Culture, Medicine and Psychiatry*, 1(1), 9-23.

Fabrega, H.:

—— 1977, 'The scope of ethnomedical science,' *Culture, Medicine and Psychiatry*,1(2), 201-228.

—— 1978, 'Ethnomedicine and medical science,' *Medical Anthropology*, 2(2), 11-24.

Fabrega, H. and Silver, D.: 1973, *Illness and Shamanistic Curing in Zinacantan: An Ethnomedical Analysis*, Stanford University Press, Stanford.

Feuerbach, L.: 1982, *L'essence du christianisme*, Jean-Pierre Osier (trans.), François Maspéro, Paris.

Finkielkraut, A.: 1996, *L'humanité perdue*, Seuil, Paris.

Foldscheid, D.: 1996, 'La médecine comme praxis: un impératif éthique fondamental,' *Laval théologique et philosophique*, 52(2), 506-507; 509.

Foucault, M.:

—— 1975a, *Surveiller et punir*. *Naissance de la prison*. TEL n°223, Gallimard, Paris.

—— 1975b, *Naissance de la clinique*, Quadrige, PUF, Paris.

Gadamer, H-G.: 1998, *Philosophie de la santé*, La grande raison, Grasset-Mollat, Paris.

Genard, J-L.: 1999, *La grammaire de la responsabilité*, Humanités, Le Cerf, Paris.

Genest, S.:

– 1981, 'Tendances actuelles de l'ethnomédecine: maladie et thérapeutique en pays mafa.' *Bulletin du séminaire d'ethnomédecine*, n°8, décembre, 5-19.

Good, B.: 1998, *Comment faire de l'anthropologie médicale? Médecine, rationalité et vécu*, Sylvette Gleize (trans.), Institut Synthélabo, Le Plessis-Robinson.

Grün, A. and Dufner, M.: 2000, *La santé, un défi spirituel*, Charles Chauvin (trans.), Sagesse, Médiaspaul, Paris/ Montréal.

Haegel, P. and Rouvillois, S.: 1995, *Corps et sagesse. Philosophie de la liturgie.* Aletheia, Fayard, Paris.

Houellebeck, M.: 1998, *Les particules élémentaires*, Flammarion, Paris.

Husserl, E. 1970, *The Crisis of European Sciences and Transcendental Phenomenology*, David Carr (trans.), Northwestern University Press, Evanston.

Ide, P.:

—— 1996, *Le corps à cœur. Essai sur le corps*. Enjeux, Saint-Paul, Versailles.

—— 1997, *Eh bien dites : don ! Petit éloge du don*. L'Emmanuel, Paris.

—— 1998a, *Mieux se connaître pour mieux s'aimer*. Fayard, Paris.

—— 1998b, 'Une éthique de l'homme comme être-de-don.' *Liberté politique*, n° 5, 29-48.

Illich, I.:

—— 1975, *Némésis médicale. L'expropriation de la santé*. Seuil, Paris.

—— 1999, L'obsession de la santé parfaite. *Le Monde diplomatique*, March, p.28.

Laplantine, F.: 1986, *Anthropologie de la maladie. Etude ethnologique des systèmes de représentations étiologiques et thérapeutiques dans la société occidentale contemporaine*. Science de l'homme, Payot, Paris.

Leonard, A.: 1980, *Pensées des hommes et foi en Jésus-Christ. Pour un discernement intellectuel chrétien*, Lethielleux/Culture et Vérité, Paris/ Namur.

Leriche, R.: 1936, 'De la santé à la maladie' *Encyclopédie française*, VI.

Lovelock, J.: 1979, *Gaia, a new Look at Life on Earth*, OUP, Oxford.

Naquet, P.: 1998, 'La culpabilité libératrice,' in M. Solemne, *Innocente culpabilité*, Dervy, Paris, pp. 57-69.

Panikkar, R.: 1998, *Entre Dieu et le cosmos. Une vision non dualiste de la réalité*, Entretiens avec Gwendoline Jarczyk. L'expérience intérieure, Albin Michel, Paris.

Pius XII:

—— 1949, Allocution à l'Assemblée Mondiale de la santé, 27 juin 1949. In: Solesmes' monks: *Ce que les papes ont dit sur la médecine*, Le Sarment-Fayard, Paris, p. 98-99.

—— 1951, Lettre à M. Charles Flory, 2 juillet 1951. In: Solesmes' monks: *Ce que les papes ont dit sur la médecine*, Le Sarment-Fayard, Paris, p. 133.

Ricoeur, P.: 1950, *Philosophie de la volonté*, Aubier-Montaigne, Paris.

Rimbaud, A.: 1972, 'Génie', in: Adam, A.,: *Œuvres complètes*. Bibliothèque de la Pléiade, Gallimard, Paris, pp. 154-155.

Saget, H.: 1976, *Mécanisme et déterminisme en physiologie contemporaine*. Nauwelaerts, Louvain-Paris.

Sartre, J-P.: 1947, 'La liberté cartésienne.' *Situations philosophiques*, TEL, n° 171, Gallimard, Paris, pp. 61-80.

Segal, H.: 1985, *Technological Utopianism in American Culture*, University of Chicago Press, Chicago.

Sfez, L.:

—— 1995, *La santé parfaite. Critique d'une nouvelle utopie*, Seuil, Paris.

—— 1997, 'L'utopie du corps parfait'. In: Bouretz, P. & Vigarello, G.: *Esprit*, 229, 43-55.

Sicard, D.:

—— 1993, 'Etre responsable pour un médecin aujourd'hui,' *Revue d'éthique et de théologie morale, Le supplément*, 186 (septembre-octobre), 118.

—— 1999. *Hippocrate et le scanner. Réflexions sur la médecine contemporaine*, DDB, Paris.

Theillier, P. and Geffard, M.: 2001, *Une nouvelle approche biomédicale des maladies chroniques. L'endothérapie multivalente*, Ecologie humaine, François-Xavier de Guibert, Paris.

Thomas, K.: 1971, *Religion and the Decline of Magic*, Peregrine Books, London.

Tourpe, E.: 2000, *Donation et consentement. Une introduction méthodologique à la métaphysique*. Donner raison, n° 8, Lessius, Bruxelles.

Verspieren, P.: 1984, 'Vie, santé et mort'. *Initiation à la pratique de la théologie*, vol. IV, Le Cerf, Paris, p. 359-401.

Wittgenstein, L.: 1971 [1914-1916], *Carnets*. Translation, introduction and notes by Gilles-Gaston Granger. Paris: Gallimard.

Woestelandt, B.: 1986, *De l'homme-cancer à l'homme-Dieu*, Dervy-Livres, Paris.

World Health Organization (WHO): 1946, 'Proceedings and Final Acts of the International Health Conference, held in New-York from 19 June to 22 July 1946,' *Off. Rec. Wld Hlth Org.*, 2, 1-143.

ARMANDO ROA

THE CONCEPT OF MENTAL HEALTH[1]

I. INTRODUCTION

It would perhaps be good to begin by briefly presenting the various ways in which mental health is defined, in order to ascertain whether and to what extent the phenomena so widespread today, such as increase in drug addiction, violence in big cities, serious instability of the family, loss of the meaning of sexuality, etc., are necessary consequences of post-modern culture, or are only the consequences of the way in which a sick society lives out its existence. The so-called post-modern culture stands alien to the inner lives of persons, and sets its goal in the faith that technology will ultimately solve all our anxieties and problems, and that life must become a permanent source of pleasure. An argument in favor of the former, that is, in favor of the hypothesis that ours is a society of fragile mental health, could be the manifest increase of conditions leading to depression and stress, such as the high level of efficiency required by the daily tasks, the merciless competition of professional life, the loneliness in which one must grow in the midst of the multitude. All of this has been documented by outstanding researchers and, furthermore, can be observed by anyone.

From an epidemiological perspective, it could be said that mental health focuses its interest on normal populations, and studies among them those groups at greater risk of becoming sick. Psychiatry, on the other hand, deals with sick individuals.

From a sociological point of view, one considers as mentally healthy those who do not show maladaptations in their habitual occupations, and do not constitute a source of discomfort for those around them. Mental diseases restrain the possibility of predicting social behavior, which is the essence of social life. In our opinion, such a point of view is limited by its exclusion of severe mental diseases like obsessive neuroses which only slightly – if at all – disturbs occupational efficiency.

The concept of "social competence" has also been considered a synonym of mental health. This concept designates the coincidence of what can be expected as appropriate performance for a person and his real

P. Taboada, K. Fedoryka Cuddeback and P. Donohue-White (eds.), Person, Society and Value: Towards a Personalist Concept of Health, 87–107.
© 2002 Kluwer Academic Publishers. Printed in Great Britain.

performance (Horwitz, 1991). This definition puts collective convenience above the individual.

It has also been said that mental health is a state of subjective well-being resulting from satisfactory psychosocial experiences (Horwitz, 1991). The problem with such a definition is, for example, that a servile person, praised by those before he bends, would be a model of mental health. Certainly nobody would agree with this.

There are other definitions of mental health, such as that of the International Preparatory Commission of the World Federation for Mental Health:

a) mental health is the state that allows the optimal development of each individual in the physical, intellectual and affective order, to the extent that is compatible with the development of other individuals.

b) It is a duty of society to allow its members to achieve that development, assuring, likewise, the development of society itself, within the tolerance due to other societies (Carstairs, 1978, pp. 7-8).

In our opinion, this definition is charged with subjective values. Thus, so called 'optimal development' would depend on what specific social groups evaluate as such.

A more accurate definition of mental health is the one proposed by Dr. Marie Jahoda:

a) awareness and acceptance of a self and a sense of one's own personal identity; b) conscious acceptance of the development and modification of one's own personality; c) integration of one's personality and the subsequent acceptance and resistance of the tensions generated by affective overload; d) autonomy, that is, the ability to make one's own decisions; e) correct perception of reality, including the "objective" sensibility towards the feelings of others; f) control of one's personal environment including capability to love, to solve problems and to act efficiently (1958, p. 8).

Finally, mental health is also frequently seen as a communal or socio-cultural characteristic describing general communal features. Thus, the quality of a community's mental health is measured according to indirect indicators: the rate of violent or delinquent behavior, alcoholism, family disintegration, etc. The same could be applied to occupational groups by measuring the levels of absenteeism, the rate of accidents, the number of sick leaves, etc. Mental health is considered as one of the quality of life indicators of a community and is part of the socio-economical development of societies (Horwitz, 1991).

In order to propose our own concept of mental health, we will sketch out in the following the common understanding of mental illness, since the way of delimiting it has naturally influenced the way in which mental health has been defined.

II. MENTAL ILLNESS: DIFFERENT MODELS

A. Biological Psychiatry

Many authors sustain the position that the method required to circumscribe mental diseases must be identical to the one utilized to circumscribe somatic diseases, that is, to investigate the physiopathological and biochemical alterations that give rise to symptoms and the clinical evolution of a specific pathological condition. Thus, symptoms are nothing but the perceptible manifestation of the disturbance of normal processes in the body, including possible genetic alterations. This concept of disease gave rise to the so-called biomedical model, and in psychiatry it is the basis of the present trend of biological psychiatry.

Nevertheless, daily experience shows something different. The concrete way in which symptoms appear and are experienced by patients, their tendency toward improvement or aggravation, and adherence to the doctor's advices, depend on other factors, such as the presence of a surrounding environment of love. Symptoms are also affected by whether the patient is in a hospital or at home, whether he is understood by the medical doctor and by other members of the medical staff, or whether familial, professional or economical problems of the moment are being taken into consideration. The classic example of the improvement or aggravation of diabetes caused by psychological factors suffices to demonstrate the point.

B. The Psychosocial Trend

The need to have in mind such factors which have become increasingly evident due to the growing depersonalization of medicine and physicians in the last decades, has turned the exclusively biomedical model of identification and treatment of illnesses necessary but not sufficient. This realization gave rise to the development of a psychosocial model for explaining the origin of mental disease. This model calls into question the possibility of identifying the origin of mental diseases from a mere biological-structural point of view. It rather places the origin of such diseases in factors arising from the dynamisms of the very society within which the individual lives.

A typical example of how this model operates is found in what happened with homosexuality. Until recently, homosexuality was clearly considered an abnormality. This notion has been discarded because contemporary society has decided that pleasure is its own justification. Therefore, it is said, as long as a homosexual finds in sex a pleasure equal to the one experienced by heterosexuals, there is no reason for considering him or her abnormal. For contemporary society, homosexual relations are perhaps only one of the many ways of experiencing sensual pleasure, and carry with them no implications about psychic health. Through such reasoning, egosintonic homosexuality has been removed from the new classifications of mental abnormalities, such as the DSM-IIIR and DSM-IV (American Psychiatric Association, 1988, 1995). Here we can clearly see the way in which cultural and social influence, and especially the active pressure exercised by the groups concerned, can change central postulates which not so long ago pertained exclusively to the realm of mental health.

C. The Antipsychiatric Trend: Szasz's Ideas

The influence of social and cultural factors on the delimitation of the concept of mental health as proposed by the psychosocial model (evident for instance in the question of possible alterations of sexual tendencies) has been developed to its extreme by some anti-psychiatrists such as Laing (1980) and Cooper (1971, 1979) as well as by Thomas Szasz (1970). Szasz, for example, has interpreted the schizophrenia of Kraepelin (1970) and Bleuler (1924) as a mere invention that allows a prejudiced society to imprison intractable individuals who protest against what they consider to be unjust. Szasz admits only slight differences between the affected language of a schizophrenic patient and the language of 'normal' persons. Proclaiming oneself to be Napoleon seems to him to be just as delirious as believing a piece of bread – the Host – to be the body of Christ. Nevertheless, the one is incarcerated as mentally sick and the other left in freedom and considered worthy of all respect. His book *Schizophrenia, the Sacred Symbol of Psychiatry* (1979) is full of similar examples, although in almost all of them there is a notable lack of clinical rigor, such as confusing a delirious apodictic idea, not based in any testimony, with a religious belief, based not in apodictic evidence but in faith.

Szasz accepts the fact that in many of the supposed schizophrenic patients one can speak of deviant behavior, but not of disease. According to him, one could not speak of disease without the actual appearance of somatic alteration. One could not rely simply on the fact that an alteration of this sort will appear in the future. With regard to neurosis, he affirms: "Freud invented neurosis in order to justify calling conversation and confession 'psychoanalysis', and in order to consider both as a form of medical therapy" (1979, p. 38).[2] In short, the social pressure to lock up individuals who are dangerous to society inadvertently led to Kraepelin's and Bleuler's invention of a nosological entity, presented as a discovery of medicine. Nevertheless, according to Szasz, this was nothing more than an articulation of a general desire to find a justifiable pretense for punishing and isolating individuals who disturb the social order. Logically, Szasz doubts that a society which is so contriving in its affirmation of the freedom of man can be considered mentally healthy.

Antipsychiatrists such as Laing and Cooper consider society in general to be abnormal. In their opinion, schizophrenic patients exaggerate this abnormality in order to redeem themselves. If society were mentally healthy, there would be no need for some persons to maximize such an abnormality, turning themselves into schizophrenics. In a certain sense, it is a necessary conclusion of these premises that in a mentally healthy society, schizophrenic psychosis would not exist.

D. The Anthropological-Phenomenological Trend

We must assert that the world of mental diseases was not constructed under the dominion either of the biomedical or of the psychosocial model. The former is reductionistic and ultimately seeks to reduce every disease to alterations in the sphere of the somatic, according to the very successful paradigm of modern science. This tendency did not exist before Galileo, Harvey or Newton, even though mental diseases had been recognized already in ancient Greek medicine and in many other different cultures (Radl, 1931; Roa, 1977). One could be led to think that these peoples used the psychosocial method ingenuously and without an explicit consciousness of it, because all, or almost all societies have distinguished the mentally ill from the delinquent, and although both of these disturb the social order, punishments were established only for the delinquent. This, however, is not such a plausible interpretation. In our opinion, this shows that it was not the failure to conform to a social order which constituted the deviation from being healthy. Moreover, one distinguishes within any social level still today, with the exception of borderline cases, the insane individual from the murderer or thief.

It is more probable that this distinction would have been established in previous ages on the basis of an *anthropological-phenomenological method*, with the difference that then it was used implicitly, and today it is used explicitly. Doubtless, before the development of the differential symptomatology which allows us to distinguish one clinical condition from another, people must have been able to observe signs that led them, intuitively, to recognize whether someone was mentally ill, somatically ill, or a criminal.

For the sake of briefness, and considering the clinical entities from the perspective of today (a point of view which must be, in some way, slightly similar to that of previous times), we would like to point out the following characteristics which indicate a mental disorder, and which can appear collectively or in isolation.

The first is the loss of the central possibility of the person to differentiate the real from the imagined, the dreamed, or the remembered. Thus, for the patient, the voices and ideas of delirium have the same degree of reality as do the real things he perceives (and sometimes even more). Confusion between dreams and waking situations occurs with relative frequency in psychoses. Now, one must remember that the difference between unreal things, dreams included, and real things (the waking world) is established in medicine only on the basis of a common-sense point of view. This point of view is fruitful in medical practice, because it keeps medicine from becoming involved in the epistemological considerations that have preoccupied the philosophical world since before

Descartes, but which have little to do with psychiatric nosological investigation.

Another central characteristic of mental disorder is the fact that mentally ill persons no longer dispose freely of their field of consciousness. On the contrary, the mentally ill person is invaded, against his will, by strange ideas, unpleasant moods for which there is no sufficient explanation, obsessive occurrences which he sees as absurd but from which he can not liberate himself, and anguishes that he cannot dispel which fill him with fear and uncertainty. All this happens without a clear reason or observable situation to explain it. Nevertheless, in the case of neurosis, the region of consciousness related to the free will to act for good or evil remains undamaged.

A third characteristic is the fact that such alterations can arise with or without the patient being aware of his sickness. This has no influence on whether or not these disturbances are cured. The anxious or obsessive neurotic patient is aware of the unmotivatedness and absurdity of his disturbances, but this does not at all diminish the torture with which these disturbances are imposed upon their victim, and there is no effort of the will capable of taking them away. The patient has had the field of his consciousness appropriated, and his self can no longer dispose of it in order to turn it to the occupations of his choice (Roa, 1981).

In the case of schizophrenic patients who hear voices and feel that others can steal or guess their thoughts, no awareness of being ill can be discovered. Patients believe that their condition is as real as the world surrounding them. They make no efforts to re-appropriate their consciousness, because in order to make this effort they would have to be able to differentiate the real from the unreal, and this ability has been lost.

In summary, in contradistinction to somatic diseases and to social deviations, such as prostitution or delinquency, in mental illness something fundamental to the person has been damaged. This is the capacity to dispose freely of consciousness, and to dedicate it, for example, to contemplation, to action, to help his fellows, control the proper direction of drives and appetites, to project the future, etc. What is also damaged, at least in severe diseases such as schizophrenia or maniac-depressive disease, is the possibility of clearly distinguishing real from imagined things, living from dead things, being from nothingness. It is the image of man in his high quality as a person which is disturbed, and which allowed primitive cultures to segregate mentally ill persons. These primitive cultures certainly felt themselves touched in an essential aspect.

In relation to all that was said above, and with regard to the often heard statement that western society is sick, we believe that this society is not lacking mental health, because this society does not confuse real with unreal things. It dedicates itself with enthusiasm and determined vision to a feverish, competitive activity, to the avid quest for money, to purely hedonistic lifestyles, even if the result of this is spiritual impoverishment, loss of the meaning of life, depression, or the need for drugs. On the contrary, because this society is healthy, it is morally responsible for the direction in which it takes its existence, for the example of conduct it gives to its members and for the consequences which all of this carries for its health.

III. MENTAL HEALTH

To define mental health means first to define concepts such as 'health' and 'mental', which are too broad to be encompassed by a concept. Let us call to mind the WHO definition of health as a state of physical, mental and social well-being, a definition which has not proven operational. If this formula were accurate, it would absurdly classify patients with incipient cancer with no clinical signs, and also maniacs or hypomaniacs, as healthy. Each of these commonly experience themselves as perfectly healthy.

In reality, health is a primary concept, which can be clarified through comparisons and certain characteristics, but not defined through broader concepts. We all immediately understand, or believe we do, what it means to be healthy. But we fall into a vicious circle if we add that this consists in not being sick, because to be sick is to lack health. Logically, the current concept of adult health involves features such as vigor, physical and mental strength, easy adaptation in the daily encounter with hostile environmental and social situations, success in professional and other daily activities, harmonious vital cycles such as those of sleeping and waking or working and resting. The tendency to project confidently immediate and long-term future is also a sign of mental health, because health is an unceasing self-realization, an intense presence of the necessities proper to the human species. Medical and laboratory tests on a subject in the above-described conditions should not show any significant deviation; such deviation would turn up as surprise, a so called 'finding'.

In the case of mental health there are, obviously, other very decisive components. Nobody doubts that good physical health is an important support of mental health. Biochemical neurotransmitter dysfunction, endocrinological dysfunctions and brain dysfunctions in Alzheimer's disease are some of the most significant examples of such a relationship. On the other hand, it is remarkable that serious diseases leading to extensive damage of the soma, such as advanced cancer, can serve as a true revelation to show that integration, strength and the acceptance of pain and death are supreme expressions of mental health, since obviously among the requirements that anyone would consider invaluable for a healthy mind: serenity, capacity of grasping the hidden positive aspects of adverse situations, and lucidity to find appropriate solutions in difficult moments.

The above-described phenomenon points to a certain autonomy of the mind in its relation to the body. The mind is what orients and configures behavior and therefore directs the body. This would not be easy to understand if the mind were dependent on any greater or lesser state of corporal vigor, and if, for example, a simple cold would leave everything at the mercy of what the body requires in such circumstances.

The capacity to sleep comfortably, or to enjoy oneself even in moments of intense stress speak on behalf of mental health. Still, the lack of these capacities does not permit us to think the opposite, if there is a certain proportion between a determinate situation and the adverse reaction to it, and if good contact with the remaining portion of reality is maintained.

Part of mental health is the recognition of personal capabilities and defects, taking advantage of the former and diminishing the latter, or trying to avoid situations in which such defects have the opportunity to emerge. It is, however, equally important to tolerate the defects of others, to focus on their virtues and to be able to avoid situations of conflict at work or at home, conflicts arising from an oversensitivity to the natural imperfection of one's neighbor. A good sign of a healthy mind is the maintenance of good interpersonal relationships at work and at home; in other words, not being what is commonly called a contentious person.

The state of "well-being" on which all or nearly all definitions insist, is not an essential part of mental health. On the contrary, it is inherent to the fortitude of mental health to bear long periods of discomfort. Conditions like those of religious asceticism or responding to the heroic demands of everyday life, or serenity at the proximity of death, are almost always expressions of good health, although such situations cannot be referred to as states of well-being, or of feeling comfortably well. However, all of this can be accepted by the person with a certain feeling of happiness, a state which basically means the satisfaction of achieving. The inner sanction of this achievement is the response which the person believes to be adequate to particular or adverse situations. The "state of well-being" considered essential by many authors leads in our opinion to confusion in the definition of health, especially mental health. It should be excluded from the essential constituents of a definition of 'health'.

A. The Anthropological-Phenomenological Concept of Mental Health

From an anthropological-phenomenological point of view, which I consider the most accurate, mental health means the integrity of the mind and the constant possibility of fulfilling what a human life in accordance with its inner nature demands for its realization in every stage of life: infancy, adolescence, youth, maturity and old age. The attempt to integrate in the best possible way the results of individual creativity with that which is given in one's cultural and material environment, to act in a free and autonomous way, also belongs to mental health. This entails openness to personal interaction with people of all ages, to giving and receiving knowledge, actions and affection and to incorporate them into the inner world that everyone must assimilate and shape in order to realize and give meaning to one's own life and to the lives of others.

The healthy mind is constituted by a self and by a field of consciousness that is free to be occupied by this self in those things that are his duty or pleasure: to reflect, to carry out a task, to solve problems, to meditate, to dream, to remember, etc. The human mind is defined by the fact that everything has to be shaped in linguistic signs. Some call this the symbolic world. An essential component of the mind is also self-awareness of its own existence and of its faculties: knowledge, will, perceptive capacity, ethical capacity. Another essential component of the mind is its receptivity to the drives and appetites arising from deep psycho-corporeal levels, as well as its capacity to order them rightly, according to what is best for the person.

On the other hand, the self is immediately experienced as psycho-corporeal. In this sense, hands, feet, the whole body, are part of the self. Therefore, an important sign of disease is depersonalization, that is, feeling the body as foreign to the self. This points to the fact that the self reveals itself as the primary component of the human person, embracing body and mind. It is possible to identify from the very beginning a masculine and feminine self, and also the self when one is a child, an adolescent and a mature adult.

The human being is interpersonal by nature. This is concretely evidenced by his finitude, contingency and fragility. The human person experiences in his innermost self the need of helping and being helped by others in all aspects of life. One requires that those things which one esteems as true, good and beautiful also be recognized as such by others. Healthy interpersonal relationships are constant, fluid, without feelings of superiority or inferiority, able to give and receive with pleasure, and always with the due respect for the inner life and freedom of each person. It is precisely the intense inner life of each person that makes it possible that one and the same reality, experienced by different persons, appears in rich and varied aspects. This makes the exchange of various perspectives a worthwhile endeavor, since it leads to the fuller realization of the whole reality. For the same reason, it is through interpersonal contact that minds are nourished. A man in absolute solitude falls into sterility and desolation. *From our point of view, to be mentally healthy is to have the possibility freely to attain the highest possible fulfillment of what is proper to human nature (one's own nature as much as the nature of those around one), making it our concern that we give the best of ourselves to others and that we receive the best of others, in an active interpersonal relationship.* The richness of the inner life is one of the highest realities we can give and receive in active interpersonal relationships.

As we pointed out before, the ability to distinguish constantly among all things worked out by one's own psyche that which is possible, impossible, probable, real, imagined, dreamed and unreal, is also a central part of mental health. It is also central to mental health to perceive one's own fragility and, for the same reason, the need to be perpetually vigilant, in order to avoid falling into error, overconfidence and ethical faults, for ethical faults deprive oneself and one's neighbor of the good. In other words, exercising a high degree of self-criticism is crucial for mental health.

IV. MENTAL HEALTH AND ETHICS

A classic mark of mental health is the ability to distinguish between good and evil, but it is necessary to point out that being mentally healthy does not imply at all that a person will readily act for the good. More precisely, mental health implies the possibility to act rightly or wrongly; thus, a man can lie, steal, or murder, assuming the responsibility and the blame for his actions, without becoming mentally ill. The psychotic patient is not responsible because he has lost the sight that is necessary for these different possibilities of acting. Obviously, mental health will achieve its maximum fullness when the person acts according to what the voice of his conscience pronounces to be just and correct. If we were to hold that mental health and goodness are the same, we would have to conclude that anyone who commits moral faults lacks health; that is, is a sick person. This conclusion would lead to serious difficulties for the concept of human freedom.

It is simply the health of the mind which allows a person to make use of his freedom. Since the temptation to evil, to acting simply for one's own benefit without taking into account the harm done to others, is always present, man must continuously strive to keep himself ethically upright, i.e., he must care for his own proper development, as well as for the proper development of others and of Nature. A mentally healthy man who perseveres in doing evil will probably put his health at risk, but will not automatically lose it because of his evil. The inner feelings of remorse and guilt that accompany him are proof that his health profoundly demands a different way of life.

V. MENTAL HEALTH AND THE STAGES OF LIFE

Undoubtedly, what is known as a healthy mind has a unique configuration at each stage of life, and if each of these configurations organically leads to the next one, the next stage will have no pathological residue or development. Therefore, receiving affection and having an adequate environment for satisfying intellectual curiosity, for satisfying the yearning to marvel at games and stories, for satisfying fantasy, for developing an ethical conscience, for building confidence in relationships with other people whether of one's own age or of other ages, are important in forming a healthy childhood across its successive stages. The same could be said about adolescence, youth, maturity and old age. Therefore, it is essential to have phenomenological knowledge about the characteristics of the affective, cognitive and volitional needs of each stage of life. This kind of knowledge is essential for establishing what a human being generally expects to achieve in his or her life throughout the different stages, as well as for establishing the extent to which having these expectations fulfilled guarantees that the new expectations for the succeeding age will also be met (Roa, 1983). Although the interactions of a person with his environment vary according to the particular way each individual perceives, imagines, feels and hopes in each stage of life, it is understood that there are certain constant, typical factors. Erik Erikson (1968) has tried to carry out epigenetic research of this kind. It may well be that his studies are distorted because he considered reality only from a psychoanalytical perspective, and did not dedicate himself to carrying out as complete and neutral a phenomenological description as possible; his effort is valuable nonetheless.

One way of protecting a society's mental health is to have the society provide ways of enabling individuals to develop all those aspects included in the definition of man specified in terms of what is typical to each particular age. As the Aristotelian point of view states, human beings are social, political, and rational beings whose existence is by nature interpersonal. In other words, human beings lack the instincts which guide their behavior according to pre-established guidelines inscribed in their nature from birth, as animals do. Therefore, humans must learn everything from the fellow creatures who surround them, that is, from their actual or foster parents. This is why humans are also considered essentially family beings.

Until his later childhood, a person does not know which food is good for him and which is harmful, and is unaware of the dangers that surround him. This is why he needs the diligent attention of his parents in order to survive. He would not receive this attention if they did not love him deeply enough to feel responsible for his life and development. Feeling protected by this love gives an individual confidence and self-assurance in the face of danger, and at the same time, allows him to recognize his value and to feel worthy of existing. If he were not loved and esteemed, he would not feel that he was worth much. The child, in turn, loves his parents because he feels they are valuable and are the most important thing around him.

Throughout childhood and with the help of the family, and later, of school and of his friendship with other children and adults, the child not only feels that he belongs to his home, but also to other human groups. He grows up giving and receiving affection and cultural and material goods, but keeping throughout the whole of this active interchange the stable balance typical of an open system in his personal individuality. During childhood and, perhaps, throughout man's whole existence, homeostasis or personal psychic equilibrium depends to a large extent on family homeostasis or equilibrium. Logically, the loss of a child's psychic equilibrium (due to biological–genetic causes, for example) disturbs psychic life of the parents. Vice-versa, quarreling parents will affect psychic life of the children. In this regard, throughout his whole life the person never reaches absolute and radical autonomy. He depends constantly on the exchange of affection, knowledge and material things with others. It is a sign of mental health that this is done fluently, without any loss of one's own inner integrity and, therefore, one's own identity. Moreover, this exchange becomes a means of guaranteeing and developing this personal identity.

VI. HEALTH AND THE PERSON

Within the person, there is nothing exclusively psychic or exclusively somatic, because both aspects are part of the whole which is the human person (Popper and Eccles, 1982). For example, it is impossible that the good functioning of the body should not be experienced as a feeling of well-being, or as a feeling of being able to dispose of the body freely for whatever one desires; and, the other way around, it is impossible that the boredom of a certain mental task would not give rise to a corporal lack of energy, to a desire to idle the time away. Therefore, one should speak in terms of predominantly corporeal or predominantly mental health, rather than in terms of a radical Cartesian dualism, which is alien to our experience. The same intrinsic relation exists between body and mind that exists between the signified and the signifier in the linguistic sign.[3] It is not easy to distinguish where one begins and the other ends, because it is in the intimate union between them that the word arises. It is impossible to observe a look, a corporal position, a movement of the hands, without seeing immediately in this a mood, a degree of mental vigor, an affective proximity or distance. Perhaps, with the exception of reflex movements, and even this is quite doubtfully, there is nothing somatic that is not intimately united with the psyche.

One easily understands the difficulty which biomedicine experiences in striving to give an account for the influence of psychic and social aspects in mental health, since biomedicine is grounded mainly on molecular biology, which is fundamentally reductionistic, in spite of its importance in the development of present-day genetics. Biological psychiatry considers psychic and social aspects of the person only as consequences of the right functioning of physiological or biochemical elements. But this is not what medical practice and everyday life show. In contradistinction, an open systems biology, such as the General Systems Theory of Ludwig von Bertalanffy (1963), seems to provide an intelligible framework for understanding how the mental health of an individual can be disturbed because of alterations in the biochemical order, in personal interactions, in the family atmosphere, or in the macrosocial order. This can happen because the individual is open to receiving all kinds of stimuli and to ordering them in accordance with his own inner integrity, unless they were of such a kind that they could debilitate or break down this integrity. In the latter case, health becomes fragile or overthrown. The family is the basic system that gives support to the individual in order to satisfy his or her need simultaneously to give and receive affection, knowledge, ethical norms, and so many other things in the course of his existence. Later, the academic and other institutions into which society is organized will give him this support.

Bertalanffy's statement that dysfunctions in any part of the biological being affect the whole *as a whole* opens the possibility of considering the human being not as a mere addition of many parts, but as a whole which is much more than the mere sum of these parts. In this way, he opens a way toward considering the human being as a person. It is singularly relevant that Bertalanffy considers persons in their essences as creators of a world of symbols, for this goes beyond pure biology in the classical understanding of this science.

Von Bertalanffy writes:

We have spoken, until now, of the organism as a system that due to the peculiarity of its organization needs research methods, laws and conceptual methods which are also characteristic. But the method is not an ideal technician that repairs a material system; it has to occupy itself with a sick man who is a psychophysical totality. Thus, an old and basic issue of philosophy is raised as a medical and clinical problem. What is man's position in Nature? What is characteristic of him regarding the rest of creation? Where do we find the reciprocal relations between body and soul, between physical and psychic disturbances? ...

Not so long ago, the issue of the unique position of man in Nature was not admitted, neither if one considered the species *homo sapiens* as a mutated form of the big monkeys, nor if according to Freud, one wanted to reduce spiritual life exclusively to biological impulses and instincts. It seems satisfactory that a change with regard to this problem has taken place. However different the points of view may be in connection with reality, they have almost without exception given man a special and undisputed position in Nature.

The problem of defining mankind cannot be accomplished in a unilateral way by natural science. I, myself, do not speak as a philosopher or a theologian, but as a biologist who attempts to characterize the special position of man. In this sense I propose the following description: the privilege of mankind, that which consists in his psychology and his behavior, is the fact that man creates a world of symbols and lives in it.

I could tell you that this definition is self-sufficient and necessary to distinguish man's behavior, language, culture and history from what belongs exclusively in the biological field ....

Leaving aside this immediate satisfaction of biological impulses, man lives in a world of symbols, not in a world of things (1963, pp. XXII-XXIII).[4]

Based on this General Systems Theory, the psychiatrist George L. Engel (1977) has attempted to revive the biopsychosocial model of medicine that was proposed by the WHO. This model intends to give a more satisfactory account of health and disease by including within these concepts biological and also psychological and sociological factors, and by appealing to the doctor's responsibility to investigate not only these biological but also psychological and sociological elements. This responsibility operates not only in a state where the presence of disease is obvious, but also in cases where it is necessary to ascertain whether a disease is present or not, regardless of whether the person feels sick or healthy. Accepting the obvious fact that there is a crisis of depersonalization in medicine (so that, due to the biologistic-reductionistic mentality in which the medical doctor is trained, everything is reduced to the use of more and more sophisticated technologies and a multiplication of laboratory tests (most of them unnecessary), Engel believes that there is an urgent need for drastic reform in medical education, a reform that must carry out a radical change in our conception of the human person. As a biological starting point for this reform, the General Systems Theory seems to him to be an effective path to follow for the achievement of this end.

General Systems Theory stands very close to the idea that considers the healthy or sick person not as a mere individual, as a repetition of an identical copy of every other individual of the same species, but as a person with unity and uniqueness. That is to say, it considers a person as an entity where the proper elements of the species are present in a unique and irrepeatable manner. 'Person' means something provided with identity and sameness. Because of this, the person always feels from within that he is the same as in previous stages of his life. This happens in spite of the already-mentioned differences that exist between the diverse ages, each of which entails entering into a crisis that must be adequately solved if it is not to become pathologically exaggerated with respect to the ideals and perspectives of the previous age. The self of childhood is the same as the self of old age (Roa, 1978). In this sense, the person is not only an individual substance of rational nature, as stated by Boethius' classic definition, but something provided with an incommunicable existence, as Richard of St. Victor sagaciously added in the XIIth century (1959). For this reason, 'person' means a being endowed with a solitary, intimate world that is reserved for the self, with an existence that is impossible to alienate or repeat, and the repetition of which would be a monstrosity.[5] This idea of existence proposed by Richard of St. Victor is different from the traditional idea which, in spite of remarkable differences, seems to have been shared by many modern philosophers, including Heidegger. Though Heidegger's view of existence has been considered to be quite different from the tradition's view, in some ways, for him as well as for his tradition, existence means being constantly directed and worried by things outside the self, being lost in worldly affairs.

According to the view of St. Victor, which we share, existence is in its depths almost impenetrable even to the person himself. It encompasses the region where great intuitions, poetic images, proposals for the future, scientific ideas, are elaborated in silence. This is what I have called the *protopsyche* in another work (Roa, 1981). The protopsyche elaborates its worlds behind consciousness, in an inner silence which is far from all urgency, with the purpose of bringing them into being in the best possible way, and of sharing results of the highest possible quality with the self and with others. Herein lies the difference between this and the Freudian subconscious, which hides from conscience and shields itself so as not to be caught by it, or turns and imposes itself upon it.

I would like to pause and consider this aspect which in my opinion is central to the mental health of the individual person or society. I believe that this aspect has not been sufficiently considered by the experts on the question. In dealing with the above mentioned inner-self, I think that it is of radical importance to keep a wise equilibrium between, on the one hand, a profound interest in looking at the inner sphere of the other (which is interesting, because looking at the other's intimacy allows us to know how others see the world and their own destiny, and to compare this vision with our own vision and to enrich ourselves in this way) and, on the other, the prudence that leads to respecting the limits of each person's inner sphere. It is necessary to keep in mind that mental health does not coincide exclusively with the mere measurable social efficiency of the person, but also with an active and creative inner life.

When we speak about the concept of person, we mean an individual being endowed with rationality and freedom to shape a cosmos according to what he receives from reality. This obviously involves other persons (past, present and future) and things. In summary, it includes what is called culture, with all its connotations and imaginable degrees of penetration. It is central to the person to move within a circle of lively interpersonal relationships and to look at reality – one's own as much as that of others – as something ever perfectible, and with a sense of responsibility to perfect or transform it. In this sense, Heidegger is right in saying that man is always in essence a project, in the sense of being directed toward an unceasing and ever-increasing fulfillment of his being.

The fact that the person, rather than impoverishing himself, strengthens his affections, knowledge and free acts the more he gives affection, knowledge and voluntary help to his fellow man, speaks in favor of mental health. As Aristotle said, if man acquires the habit of doing the good, it will be easier for him to act for the good. On the other hand, it is obvious that one who loves more generates more love in the depth of himself. This is shown by daily experience. It is proper to mental health that this exchange of knowledge, affections, acts and work is easily accomplished. This engenders a state of inner serenity, of peace. Now, the problem of true health, and in particular of mental health, which allows every person to be faithful to what is demanded by his own nature, is that it is necessary to know how to give and receive all of these things. Normally, one learns to do so through education, example and counseling. Affections, knowledge and acts indiscriminately given, without paying attention to the particular way in which the other person lives out his existence, could be harmful and not at all beneficial, as can happen in an over-protected childhood. On the other hand, one understands that to give excessive amounts of knowledge, with the demand that it should be assimilated and cultivated, could provoke inhibitions and feelings of inferiority, analogous to those provoked by excessive punishment, lack of affection or the relegation of the other to a secondary position. All this points to the fact that each person has his own peculiarities in giving and receiving, because he is unique and has an incommunicable existence. If we want to act for the good of each person, for his mental health, we must study such peculiarities.

It is commonly known how much a disappointment, an obscure horizon for the fulfillment of life projects, the loss of a desired situation, a broken marriage, the death of a beloved person, or being the victim of undeserved suspicions, can influence the development of illness. These situations involve the whole organism, starting with the immune system. In order to recover health, intense affection and esteem from those who surround the person is required, and especially from the medical doctor to whom the person comes looking for help. This help cannot be reduced to the use of sophisticated technologies and medication. It must return to the person his faith in existence and brighten in some other way his destiny. For that, it is necessary to give him love, to make him feel that we share intimately in his problems, and that we are sincerely interested in finding a solution. For all this, it is necessary, as we have already mentioned, to know his personal way of being, to know the way he is able to assimilate affection and suitable assistance. This has its own peculiarities in each person. This is why some of the stereotypical gestures of a physician, such as sympathetic smiles, handshakes, and stereotypical kind words are not well received by so many patients; they are rejected when it is realized that these gestures are not directed to their individual inner self.

The ways of carrying out such a therapy are being investigated today, inasmuch as we have liberated ourselves from reductionistic theories like those of the old molecular biology or of psychoanalysis. Now we are beginning to ask ourselves what the human reality is, and to consider that a person's mental health will become stronger as he succeeds in becoming the faithful reflection of the essence of his nature – that is, to have a unique position in the world of living beings: that of being a person and belonging to a society of persons.

Finally, we may say that health constitutes what everyone would wish to have because it enables us to be identified as human beings, to freely develop everything that one carries within as a seed and to open in a creative way to other minds. Therefore, referring to mental health and referring to the essence of mind are different modes of pointing to the same thing, and describing one is describing the other. Illness consists in the successive or simultaneous negation of the various properties which a mind must possess in order to fully deserve such name. Consequently, health is not the mere negation of illness, but just the opposite: health is what enables the mind to assume its completeness. To define briefly what mental health is means to define in brief what is encompassed in the essence of the concept of mind. This is not an easy task. Perhaps it can only be achieved through successive elucidations, of the kind we have attempted to provide here.

Universidad de Chile
Santiago, Chile

NOTES

[1]   This text was translated from the original in Spanish by Paulina Taboada.

[2]   This quote was translated from the Spanish text by Paulina Taboada.

[3]   In a linguistic sign, the signified is that which is named by the sign, and the signifier is the verbal or written symbol by means of which it is named.

[4]   There is no official English translation of the Preface of Bertalanffy's 1963. This quotation was translated from the Spanish text by Paulina Taboada.

[5]   Richard of St. Victor (1959, p. 243) states: "The word person must be understood as an individual, singular, non-communicable property. Thus I think that the considerable difference between person and substance will be easily recognized" (Translated from the French text by Paulina Taboada).

REFERENCES

American Psychiatric Association.: 1988, *Manual diagnóstico y estadístico de los trastornos mentales*, Masson, S.A., Madrid, Spain.

—— 1995, *Manual diagnóstico y estadístico de los trastornos mentales*, Masson, S.A., Madrid, Spain.

Aristóteles.: 1988, *Etica Nicomaquea*, Gredos, Madrid.

Bertalanffy, L.v.: 1963, 'Lecture presented in Ratisbone, in 1961, at the Solemn Jubilee celebrated for the Opening of the 25th Course on Medical Studies,' in *La concepción biológica del cosmos*, Dr. Faustino Cordon (trans.), Ediciones de la Universidad de Chile, Santiago.

Bleuler, E.: 1924, *Tratado de Psiquiatría*, José M. Villaverde (trans.), Prólogo Santiago Ramón y Cajal, Calpe, Madrid.

Carstairs, GM.: 1973, '¿Qué es la salud mental?,' *Salud Mundial*, Mayo, 4-9.

Cooper, D.: 1971, *Psiquiatría y Antipsiquiatría*, Jorge Piatigorsky (trans.), Paidos, Buenos Aires.

—— 1979, *El lenguaje de la locura*, Alicia Ramón García (trans.), Ariel, Barcelona.

Engel G.L.: 1977, 'The need of a new medical model: A challenge for biomedicine,' *Science* 196(4286), 129-135.

Erikson, E.: 1968, *Identity, Youth, and Crisis*, A.W. Norton, New York.

Horowitz, N.: 1991, 'Una visión de la salud mental desde la sociología,' *Rev Psiquiatría* VIII, 859-868.

Jahoda, M.: 1958, *Current Concepts of Positive Mental Health*, New York, Basic Books.

Kraepelin, E.: 1970, *Lecons cliniques sur la démence précoce et la psychose maniaco-dépressive. Textes choisis et présentés par Jacques Postel*, Edouard Privat, Toulouse.

Laing, R.D.: 1971, *Experiencia y Alienación en la Vida Contemporánea*, Paidos, Inés Hülze (trans.), Buenos Aires.

—— 1980, *Los locos y los cuerdos*, Silvia Furio (trans.), Crítica, Grupo Editorial Grijalbo, Barcelona.

Popper, K. y Eccles, J.: 1982, *El yo y su cerebro*, C. Solís Santos (trans.), Labor Universitaria, Barcelona.

Radl, E.: 1931, *Historia de las Teorías Biológicas*, Félix Diez Mateos (trans.), 2 vols., Revista de Occidente, Madrid.

Roa, A.: 1977, *Algunos momentos señalados de la historia de la medicina a través de las variaciones en la visión del cuerpo humano* (unpublished manuscript).

—— 1978, 'Las edades de la vida y el sentido de la madurez avanzada para una nueva civilización,' *Revista de Psiquiatría Clínica* XXIV, 5-18, Chile.

—— 1981, *Psiquiatría*, Andrés Bello, Santiago.

—— 1983, *El mundo del Adolescente*, Editorial Universitaria, Santiago de Chile.

Saint Victor, R.: 1959, *La Trinité*, transl. from Latin text by Gastón Salet, S.J., Les Editions du Cerf, París.

Szasz, T.: 1970, *Ideología y Enfermedad*, Leando Wolfson (trans.), Amorrortu, Buenos Aires.

—— 1979, *Esquizofrenia, el símbolo sagrado de la* Psiquiatría, Mercedes Benet (trans.), Premia, México.

JOSEF SEIFERT

WHAT IS HUMAN HEALTH? TOWARDS
UNDERSTANDING ITS PERSONALIST DIMENSIONS

I. INTRODUCTION

Health is, in the final analysis, something ultimate, in the sense of being
an irreducible datum which is only found in living things, and must be
understood in its own ultimately undefinable terms. This is *a fortiori* true
of that quite unique phenomenon within the more general datum of health
which we can call "the health of a person." "Personal health" is an
entirely new datum which cannot even be reduced to what we call
"health" in plants and animals. Using an expression in Heidegger's *Sein
und Zeit* (1927, § 7), a primary *phenomenon* can never be defined through
other elements but can only "show itself from itself." This is true even
though health, and specifically the health of persons, is not as ultimate a
datum as life or personhood, but depends on them, as well as on some
pre-biological elements. The latter explains the partial applicability of the
"biomedical model" to health and the practical fruitfulness of various
forms of reductionism. Such reductionist theories take note of isolated
moments of health which are truly elements of it, though any reduction of
health to them is untenable.

Something similar is true about the pair of concepts in terms of which
Nordenfelt seeks to interpret health: goals and capabilities. While these
terms take into consideration the biological and conscious levels of
human life, and hence allow for a less reductionist concept of human
health than the purely biomedical model, they concern one element that is
merely a *part* of health (capabilities to accomplish goals) and another one
which is neither indispensable to nor sufficient for the presence of health:
goals (at least when interpreted in the sense of consciously adopted
goals). Such an action-oriented concept of health could not explain why
the cancerous person who is presently still capable of accomplishing all
of his goals is nevertheless unhealthy, nor could it explain why a person
attracted by a *dolce far' niente* in which he does not have or pursue any
goals is healthy. Also, this action-theoretical account of health is
insufficient to capture both the deeper essence and all the elements of
health in general, and of the health of persons in particular.

*P. Taboada, K. Fedoryka Cuddeback and P. Donohue-White (eds.), Person, Society and Value:
Towards a Personalist Concept of Health,* 109–143.
© 2002 *Kluwer Academic Publishers. Printed in Great Britain.*

The nature of health, however, must not only be understood in light of all its parts, but also in light of other more primary data such as life, rationality, and personhood. Within the realm of irreducible data, while all of them are ultimate in a broader sense of this term, some are still more ultimate and primary than others. For example, being is entirely ultimate, life already presupposes being; color is more ultimate than 'red' which is an irreducible species of it but includes the more general moment of 'color' in its essence, etc. In this way, health must be conceived in the light of more primary data such as life. Nevertheless, just as the color 'red' cannot be reduced to the notion of color as such or to any other color, let alone wavelength, so 'health' can in no way be reduced to life or to any other more general and more primary datum. Accordingly, not only mental health, but also those elements of human health which reductionist theories of health recognize, will never be properly understood if they are not seen in their specific personal dimensions. This paper undertakes some effort to demonstrate this point. More specifically, I shall attempt an elucidation of four facts:

1. All the pre-biological aspects of health receive a radically new character and meaning when they are part of the health of persons.
2. All the specifically *biological* aspects of health which depend on the primary datum of *life* (as it is found also in plants and animals) receive an entirely new personalist meaning and new characteristics when they exist in human persons.
3. There are many uniquely personal dimensions of health, particularly of mental health, which do not exist in plants and animals at all.[1]
4. Personal health, while it constitutes a basic human good, neither constitutes the absolute and highest value in human life nor can it be realized fully without reference to higher moral and social values.

II. THE ROLE OF THE CONCEPTS OF SPECIES-PLAN, NATURE, DYNAMICALLY DEVELOPING TEMPORAL PATTERNS, AND GENDER IN AN UNDERSTANDING OF HUMAN HEALTH

Speaking quite generally, the health of any organism is related to the specific nature of the given living being, which dictates the range of what constitutes its health. In order to know which abilities belong to the health of a cat or a dog, I must presuppose some knowledge of the respective natures of cats and dogs. Understanding human health implies as well an understanding of the specific nature of the human being. To have insisted on the central role of a species-plan for understanding health is a significant contribution Boorse (1975; 1977) has made toward an appropriate understanding of health. Without understanding the natures, tendencies, intended actualizations and species-plans of living beings, the notions of health and specific health of a given being cannot be understood.

This fact is to some extent even recognized by those empiricist theories which abandon the notion of nature or essence entirely, reducing them to mere statistical standards. Boorse (1975; 1977) himself proposes such a theory, notwithstanding some Aristotelian elements of his philosophy. He seeks to replace the notion of nature by "normal function" and defines "normal function" in reference to the species (for example, man) as reference-class. The "species-design" (still a reminder of an Aristotelian concept of nature) is then reduced to a statistical quantity, although not entirely, in that Boorse permits over-function with respect to the species design and other elements which are not explainable in purely statistical terms.[2]

But is it not evident that a statistical quantity can never replace the notion of nature as a guiding principle of the concept of health? If we all had cancer, AIDS, headaches, missing limbs, or were blind, this would not make such states of statistical normalities "healthy." Thus it is clear that we have to look beyond facts and statistics to comprehend human nature as a standard for what is healthy and unhealthy, as Nordenfelt (1986) also points out. An in-depth study of the problem of essences would show that not only necessary essences but also meaningful contingent natures are in no way reducible to mere statistical generalities, and require categories and methods which the "negative Popperian empiricism" of the theory of falsification also cannot account for (Hildebrand, 1960; Seifert, 1991, 1995b). This applies much more to the nature of the human person which involves both highly meaningful contingent structures and essential necessities, without which particularly mental health remains incomprehensible.

Especially in light of nature as the second underlying moment of 'health', it becomes evident that some kind of "biopsychosocial model of health," for which Engel (1977) argues, is needed today. But even this model needs to undergo a critical examination and particular personalist reform, the adequacy of which will depend on the adequacy of our grasp of human nature. For this "health-model," which corresponds to the WHO definition of health, could be interpreted both in an unjustified subjectivist sense which denies the objective nature of the human being and of his or her mental acts, and in a utopian sense which would include all physical, psychic, and social goods of the human person under the umbrella of 'health'. Such a concept is both inadequate and useless for medicine, as we shall show when discussing the concept of 'mental health' later in this paper.

Besides nature, a distinct dynamic component of health also becomes evident at this point. What would be health for a baby would be a terrible retardation for an adult. As Engelhardt (1981, p. 41) notes: "The acceptable physical state of an 80-year-old would be disease for a 20-year-old."

The formation of an adequate notion of human health also requires a grasp of the gender-related differences between human beings and of the dynamic development within a given nature. What would be normal physical and psychic states for a menstruating girl would be abnormal for a young man, etc. Of course, some of these "normalities" may themselves be sicknesses, such as old age and its effects themselves, which the Romans call *morbus* (disease). Other cases may still be normal and meaningful when the rhythms and gestalt-quality of the entirety of human life is taken into consideration – for example, though it involves the loss of a fundamental dimension of her generative power, the infertility of a post-menopausal woman is normal. This again shows the need for an intuitive grasp of the form and essence of the human person (including the form of dynamic development) as standard for human health. And this is precisely the task I undertake in the following sections of this paper.

III. THE PRE-BIOLOGICAL ASPECTS OF HEALTH AND THEIR
PERSONALIST DIMENSIONS

There are at least three aspects of the health of organisms, and in particular of human health, which are "pre-biological" in the sense that, while they are parts of health only in living things, these elements can also be found in machines or statues or other non-living objects. For this reason, certain reductionist models of health (e.g., mechanical or chemical models) can be successful insofar as they concentrate on these pre-biological aspects and analyze or "fix" them.

A. The Mechanical and Biochemical Functioning of the Body

The first of these pre-biological elements of health is the inner chemical and physical order and the physico-chemical, electrical, and mechanical functioning of the body. While these aspects of health partake in the structures of organic chemistry and in specific functions of the organs of living beings (and thus are never reducible to inorganic chemical occurrences or to the mechanics of a machine or camera), they nevertheless contain, on a certain level, strict parallels to the chemical processes, mechanical functionings or optical devices which can also be found in lifeless things and machines. As in lifeless things, in living organisms chemical deficiencies can be treated by injecting the necessary chemical substances, and the purely mechanical and optical aspects of vision or hearing may be repaired by surgery or by new lenses.

Yet any purely mechanistic consideration of these pre-biological elements of health would be mistaken. This applies particularly when these material functions also affect the specifically human body-soul relationships. The breakdown of the mechanics of the inner ear of a *human* being is not an isolated material or physiological dysfunction, but the cause of deafness and problems of health which can never be explained in terms of a mechanistic philosophy of man. The breakdown of the mechanical functions of the inner ear may even give rise to the despair that was expressed vividly, for example, in Beethoven's *Heiligenstädt Testament*.

To understand the pre-biological elements of health as part of *human* health requires therefore the understanding of the specific *medial role* of the mechanics of the ear for the experience of hearing and for the higher sphere of intentional conscious relations of persons to music and language, and to the entire aesthetic and cultural world with which perception brings us into contact and which perception renders possible in persons. The same applies, for example, to the mechanical functioning of the joints, without which humans would lack the capacity to perform many acts.

The link between these pre-biological conditions and elements of health and the person is not restricted to the *conscious* relationship between the personal life of the human person and the body. It involves also many unconscious relations between mind and body, and a purely ontological relationship between the human person and the body – a relationship which allows us to define man as a person-in-a-body (Seifert, 1989, 1995b). Nevertheless, the full character of the first sphere of pre-biological elements of health as parts of personal health reveals itself only in light of the relationship between them and personal consciousness.

B. The Integrity of the Form of the Body

Something similar is true of another pre-biological and aesthetic aspect of health, namely the integrity of the form of the body. Some basic parts of this integrity of the form of the bodily parts and limbs is not only a necessary requirement for the mechanical aspects of health but is also a part of health in its own right, at least if we take health in the broader sense of the fundamental bodily good whose promotion constitutes one main object of medicine. One could distinguish a narrower sense of health (the opposite of which is the "unhealth" caused by disease, sickness, or illness) from a wider sense of health that includes well-formedness and freedom from injury as well as basic feelings of wellness. Following this distinction, when taking the term "health" in its narrower sense, one could list health, integrity of bodily form, and freedom from suffering as three different bodily and psychological goods which constitute different important ends of medicine.

But in this paper, I am using a broader and still legitimate sense of health as the more encompassing good of the body which is contrasted with and also affected by injuries, malformations, mutilations, and any form of ugliness which endangers the integrity of the human form. Medical practice certainly confirms the significance of this pre-biological aspect of health. Think of the malformations of embryos caused by German measles during the first three months of pregnancy. The resulting deformity is so severe that many physicians recommend therapeutic abortion[3] – and this recommendation, however immoral it is, underlines the significance of the integrity of well-formedness for health. Or think of the practice of correcting irregularities in tooth alignment, operating on the mouth or jaw, or fitting artificial dentures, which are performed not because disorders in the mouth may cause diseases in other parts of the body, but merely for the reason of the integrity of the form and appearance of the human face (DeBakey, 1995). A whole class of diseases affects primarily the appearance of the body, such as various diseases of the skin (Lederer, 1995).

Of all branches of medicine that occupy themselves with this part of health, plastic surgery is the most prominent, dealing with the remodeling of any portion of the human body that has been damaged or deformed, be it congenitally or the result of injury or deforming surgery required in treating diseases such as cancer. The fact that facial surgery requires artistic as well as technical skills likewise shows the significance of this aesthetic aspect of health.

Plastic surgery today is often done for purely cosmetic reasons, to remove blemishes or change contours which have nothing to do with a lack of health. However, I consider here only the very fundamental aesthetic dimension which touches the integrity of the human form, which may be restored by plastic surgery, for example, when cartilage is taken from a patient's rib and then sculpted to reproduce the shape of a missing ear.[4] We must distinguish between the basic integrity of the human body and more refined aesthetic values on which individuals disagree.

Well-formedness is an aspect of health that is pre-biological because it can also be attributed to a statue or image of a body. We speak here not only of a neutral or purely functionally interpreted integrity (wholeness) of form and of useful or necessary parts, but also of an elementary aesthetic aspect common to the well-formed human body and the well-formed statue.

On the one hand, the integrity of the human body, while it involves a basic aesthetic value (given that the form of the human body as such is recognized to possess a high aesthetic value), has a human significance which cannot be reduced either to that of the aesthetic value it bears or to the usefulness of the integrity of the body for performing bodily actions. It possesses a certain ontological value in itself, that of being in conformity with the nature and species-plan of the human being. On the other hand, integrity does possess a basic aesthetic value, at least if one takes it for granted that the form (species-plan) itself of whose integrity we are speaking possesses an aesthetic value. (Think of the crucial role of *integritas formae* as one of the elements of beauty next to *consonantia* and others, according to medieval aesthetics.) While I cannot here give reasons and arguments to prove the objectivity of beauty and aesthetic value in general (Hildebrand, 1977, Seifert, 1990, 1992), I do insist on this basic objective aesthetic value as an important aspect of the bodily good and even of health in the wider sense of the term.

Moreover, higher beauty definitely falls outside of the scope of health even if we assume here the full objectivity of beauty and higher aesthetic values. One could even be tempted, in view of this difference, to juxtapose integrity of the human form and beauty (understanding under the latter only aesthetic values which go beyond those that are attached to the integrity of the human form). But I would like to insist on both a basic aesthetic value and an integrity of the human body as significant pre-biological aspects of health. The grave problem of health that results from, or even consists in, malformation throws into relief the crucial part of health which lies in a fundamental integrity, wholeness and specificity[5] of the form of the human body. This part of health does not derive exclusively from the fact that human organisms cannot "function" well in a state of malformation, but is also rooted in the elementary violation of the aesthetic integrity of the form of the human body which, in its orderly structure and beauty, can be compared to a great masterpiece of art and surpasses it in many respects, as Kant noted (1793).

In addition, this pre-biological aspect of health takes on an entirely new role in the *human* body, and this for various reasons. In the first place, the mere fact that a given malformation disfigures a *person's* body makes it very different in nature and significance from the case in which an animal body is malformed, because the nature and dignity of the person endows his body with the higher nature of a person's body and imbues it with another and higher dignity.

Secondly, the malformation of a human body is related in various ways to human consciousness and this fact makes it quite different from the malformations of an animal's body. The very fact that these malform-ations are the object of human consciousness, both of the consciousness of the subject of such malformations and of the consciousness of others (these in turn becoming the object of the consciousness of the subject of malformations), makes them very different from similar malformations in animals. One may speak here of a psychological dimension of this aspect of health which lies in the well-formation of the body.

For this reason, for example, the crippled hand of a human child is a specifically human opposite to health (in the wider sense of this term); this is quite distinct from a similar defect in an animal. Imagine, for example, a donkey who looks like a horse! Since he would neither notice nor care about the unfittingness or the change of his appearance, he might still be a perfectly healthy donkey who just looks like a horse. But a human person who is afflicted with the deformities described above would for this very reason be affected in his or her *personal* health.

Thirdly, the malformations of the body take on an essentially different character in humans because the human body is an instrument of mental acts and therefore any destruction of the basic integrity of form affects the entirety of human life, of mental and spiritual acts, of relations to others and of happiness.

C. The Expression of the Person's Essence in the Body

Closely related to the second pre-biological element of health is a third one that we could call the "bodily expression of the metaphysical nature and conscious life of the human person." The body is the visible appearance of the invisible but intelligible nature of man. The dignity of the person expresses itself in the body, in the formation of the human hands, in the upright posture or in the nobility of the form of the human head and forehead.[6] Therefore, any animal-like formation of the human face or the human hands (for example in the rare defects, sometimes seen in museums, which cause human hands to look like animal paws) strikes us as a very serious opposite to a person's health. Something similar is true of relatively frequent malformations of the head or hands of babies. A surreal thought experiment can help us to understand this even more clearly. A human child who looked even faintly similar to a pig would be the object of profound compassion, and if someone were to declare him perfectly healthy because of the well-functioning of his body, the parents of such a malformed child would regard this as a mockery. And so would the child. Consider also the horrible human countenances, mixed of ugly and inhuman forms, that Michelangelo painted of Mino and other diabolical faces in his *Last Judgment*. Instead of expressing the dignity of the human person, they express animal-like and demonic features.

We can say more trivially and generally: Any human face or human body which is disfigured by such distortion, mutilation, anamorphosis, deformation, grimace, misproportion, or misshapenness that it contradicts the visible and audible expression of the dignity of the human person in the body, reduces, by the very fact of these malformations, health and an important part of the physical well-being of the person.

Also, however, this pre-biological and quasi-aesthetic element of a human person's health requires a grasp of the essence of the person. It is a pre-biological aspect of health and a dimension of the species-plan of the human being which is by its very nature (as a metaphysical expression of the dignity of the human person) a personalist dimension of human health. When it is impaired, not only the "non-specificity of the human form" (which constitutes an opposite to the second pre-biological aspect of health) strikes us here, as it would strike us in an animal, but also the objective disproportion between the rational nature and dignity of the human person and the impairment, disfigured state or partial destruction of its visible and audible expression in the appearance or voice of a given human body.

Apart from the purely objective aspect of this element of human health, this third pre-biological aspect of health, the expression of the person's essence in the body, is also closely related to human consciousness. Therefore, grave mental and personal consequences, such as profound disturbances of relationships, self-esteem, or one's self-perception follow from such malformations, which adversely affect the personal aspect of the human body. Human health therefore regards also the totality of physical, psychic, and spiritual factors which are interwoven in human experience with such malformations or with the normal human aspect.

Preceding these psychological and social consequences of the absence of the expression of the dignity of personhood in the human body, however, there is an intrinsic disorder in its absence which constitutes a lack of health. Therefore, I cannot agree with Leon Kass (1981) when he seeks to exclude all aesthetic and even mental and social aspects, conditions and immediate consequences of a person's health, although I sympathize with his attempt to avoid the "totalitarian" elements of the WHO definition of health (see also Pjörn, 1984).

It is crucial to distinguish those dimensions of aesthetic beauty and the perfection of the physical expression of the rational and spiritual nature of the person that fall beyond the realm of health from those that constitute an important part of health. Health, in the wider sense of this term distinguished above, includes only the *basic intactness* of these aspects: (1) a *basic well-functioning* of the senses and of the body-machine, (2) the species-specific form of the human body (in contrast to its higher beauty), and (3) a fundamental metaphysical expression of the dignity of the person in the visible and audible aspects of the human body.

Someone might object to talking of "pre-biological" elements of health[7] and declare such an expression "intrinsically contradictory" because health always includes reference to life. In reply to this objection we can say: While the fact that these elements of health can exist in non-living things makes them "pre-biological parts" of health, these aspects of health also become elements of health only by their reference to life, and they become elements of personal health only in virtue of their relationship to personal life. These "pre-biological elements of health," however, are not in their own nature related to organic life, as, for example, problems of growth are. "Pre-biological" then does not mean that *as parts of health* they are not related to life. It means that in their own nature they are not related to the world of the life of the organism, but rather to chemical, mechanical, or aesthetic aspects of the body. Moreover, nothing forbids us from finding in a certain being some aspects or qualities that do not directly and as such partake in its own nature. The famous distinction between the *actus humanus* (the specifically human act) and the *actus hominis* (activities *of* human beings which are not specifically *human activities*, such as sneezing) and many other examples teach us this. In another sense, this fact that a living being possesses qualities which are not essentially related to life applies also to the way in which a being is a person in contradistinction to the way in which he or she is a being and is subject to first principles, such as the principle of contradiction. Being as such and the principle of contradiction do not refer to a human being *as human being* (*inquantum homo*) but to him *as a being* (*inquantum ens*), i.e., as possessing a character which he shares with all other entities. The difference between these two kinds of "pre-biological" elements of life and health, however, lies in the virtually identical way (which gives rise to univocal terms) in which chemical properties exist in living organisms and in the human body, versus the analogous way in which all things are beings without possessing this character of being in a fundamentally identical or univocal sense. Under "pre-biological" characteristics of health I have in mind the first of these two ways in which living beings and their health possess properties of the non-living world.

The recognition of these three pre-biological aspects of human health and their specific personal ramifications reconfirms the WHO definition's affirmation of aspects of mental and social well-being as parts of human health. We should, however, nevertheless reject the utopian elements of the WHO definition and of any definition of health in terms of a "*complete* physical, mental, and social well-being" which clearly identifies health with a good that falls beyond the scope of health. Rejecting with Kass such an illegitimate extension of the wider sense of human health, we nevertheless reject the conclusions Kass draws, specifically his rejecting entirely the second and third aspects of the pre-biological dimensions of health and ignoring the specifically personal implications and dimensions of human health linked to these three pre-biological aspects of health.

The concept of personal health I am unfolding here is also entirely free of subjectivism and relativism. In the first place, I fully admit those objective factors which also make entirely unconscious events occurring in the body (whether or not they perform the role of unconscious conditions, causes or effects of conscious acts) parts of the human person's health. Second, human conscious life, while being the life of a subject (i.e., of a person), is just as objectively real and possesses just as much of an objective existence and essence as does any non-personal being. The person in his or her conscious life is even the climax and archetype of objective reality. Third, while most of the specifically *personal* dimensions of human health are necessarily correlated with human consciousness, the biological health of the human organism as such does not require human consciousness, and in this sense diseases or disturbances (e.g., of the inner ear) remain parts of unhealth even in a person who is in a permanent vegetative state. Fourth, human consciousness itself and the different conscious acts and faculties as well as their relationship to health possess fully objective essences.

IV. PERSONALIST DIMENSIONS OF BIOLOGICAL HEALTH

Health, while it possesses the described pre-biological elements, can properly exist only in living things. Non-living beings can merely display properties analogous to health. But neither such analogies of health nor artificial simulations of life or health should be confused with these phenomena themselves. As Spaemann (1994) points out, we witness today a great ability to simulate both life and thinking, and then forget that we are dealing here with mere simulations of life and thought.[8]

This sheds further light on the sense in which we spoke of "pre-biological aspects of health." We meant by this term those features of health which, *as such*, can also exist in non-living beings, in contrast to those aspects of health to which we now turn and which constitute qualities and perfections that exist exclusively in living beings. We did *not* mean, however, that pre-biological aspects of health constitute health in non-living beings. Only living beings can be healthy, and therefore those features of health which directly touch on the essential characteristics of life are more profoundly connected with health than those which do not, for health partakes in all essential traits of life. Biological life is, empirically speaking and referring to obvious notes of organic life, characterized by nutrition, growth (morphogenesis), regeneration, propagation, endogen motility, adaptativeness, etc. With respect to all these essential features of life, we can also speak of health when the respective qualities, or at least the capacity for them, exist in a flourishing, non-inhibited form.

Can we only refer to empirically given distinct properties of organisms or is there, as the measure and standard of health, a deeper unified essential core[9] of biological life, such as the *entelechy* of which Aristotle (in *De Anima*) and Driesch (1928) have spoken, which implies the dynamic and teleological "building itself up" of the organism? That both biological life and biological health can only be conceived properly in terms of such a deeper core of biological life is suggested by the fact that none of the mentioned single features of life is absolutely necessary for biological life or health. Many organisms that no longer grow, that are sterile and unable to propagate, live and are even healthy. Even nutrition can, for an extended period of time, stop, while life and health continue.[10] Thus biological life is an actuality that lies deeper than the phenomena in which it is manifested.

But while life can continue to exist when these empirical properties are not actually present, and is a necessary *condition of health,* life – inasmuch as it transcends the mere functioning of the body and does not even have to possess in actuality many of the essential features of life such as fertility or propagation – is not the clue to health. Health precisely implies more than life as such, namely the living entity's ability to actualize its life in accordance with its nature. The standard of health is precisely the full presence of the characteristics of life, at least in potency.

There is a further point to consider: The fundamental characteristics of biological life as such are usually also characteristics which exist in specifically different modes in different living beings. To determine their health, we must therefore also grasp the marks of their species, as Boorse (1977) has pointed out. This applies particularly to the human person whose health demands much more than bare life: namely, the capacity of actualizing this life, and hence also the capacities and faculties connected with life in its specifically human form. This can be better grasped after recognizing that health constitutes a special sphere between act and potency, and involves in particular potentialities regarding the characteristics of life and the given species of life to which the given entity belongs. We will investigate in the following the characteristics of health of persons that are related to the essential marks of biological life, and investigate three personalist dimensions of biological health.

A. Self-motion, Self-generation, and Other Essential Characteristics of Bios-life in Their Literal and Analogous Aspect, as Clue to a Personalist Concept of Health

Some philosophers identified the innermost property of life as self-motion, which Plato emphasized in the *Phaedrus* (245 c 5 ff.). Also Kant characterized the living being in terms of "the inner power to determine itself to act from an inner principle" (1968, p. 544). While we agree with Bonaventure (*Opera Omnia*, vol. II, p. 744 arg. 7) who insists, "Life is prior to motion," we have recognized that health is not based on the bare naked reality of life but on its fuller, or at least its normal, flourishing.

While life also includes passive and receptive forms of activity, self-motion is an important mark of life. Therefore the actual ability of self-motion is of the essence of health which consists, with respect to this mark of life, in the actuality and most of all in the concrete capacity for self-movement, an important aspect of health which is threatened by all kinds of paralyses or hindrances to self-motion.

Also here, the absolute necessity of conceiving human health in terms of categories which are new with respect to organic and animal life as such, and which do justice to the mode in which self-motion is characteristic of a person's life, emerges clearly before our minds.[11] For if a person is reduced to the type of self-motion found in vegetative life, the permanent vegetative state,[12] which is enough for the self-motion that is part of the health of a growing head of lettuce, this constitutes a horrible state of unhealth for the human being.

The phenomenon of "self-motion" is a typical case of an analogous property of life: that is, different forms of "self-motion" are similar while being utterly dissimilar in kind. The phenomenon of a living entity "moving itself" appears in radically different ways in nature, starting from the lowest level of the *Eigenbewegung* and the capacity of adaptation of plants up to the spontaneous free acts by which a personal subject is the conscious source of his acts. Aristotle asserts in his *Eudemian Ethics* that man "is the lord over the existence or non-existence (of his actions)" (1223 a). Cicero is equally clear on freedom in his thesis that it is evident that man is an absolute origin of his free actions and that this is the necessary condition of the entire moral and legal order (Cicero, *De natura deorum* II, 44, *De officiis* I, 70,[13] *De fato*). This contradicts the common prejudice that ancient philosophy did not recognize freedom in its full metaphysical implications.[14]

In light of a philosophy of freedom, we recognize that a personalist concept of health does not merely depend on a general notion of self-motion as characteristic of all biological life, but rather on the grasp of the unique form of self-motion proper to persons: freedom, which was recognized already by the ancients and particularly by Kant as an archetype of self-motion. Self-motion, receiving a radically new character in freedom, cannot be understood as part of the health of persons if we prescind from this specifically personal form of self-motion to which the mere vegetative forms of self-motion are only analogous. Because *free self-motion* belongs to the nature of personal human life, human health involves the ability of genuinely *free* self-motion, and not only of the self-motion characteristic of vegetative and animal life.

Health is also linked to another moment of life, namely that of self-generation and, in a certain sense, the "self-creation" of a living being. Like self-motion, self-generation is a concept that refers to fundamentally different levels and spheres of life. Even when solely applied to the *bios*-life of the organism, it refers to fundamentally different data: in nutrition, the organism produces itself constantly anew. This applies also to the growth in which the organism builds up its form from a seed or a tiny fertilized egg or embryo. Another dimension of the self-generation of the organism takes place in the regeneration of old or lost cells, the healing of wounds, the restoration of health, or the regeneration of torn flesh or broken bones. Here a meaningful whole is restored. In procreation this feature of organic life extends to other individual organisms (Cf. Kant, 1793, p. 389). All these phenomena of self-generation can be healthy or afflicted by disease.

Again, the need for a properly personalist understanding of human health reasserts itself on the level of this feature of biological life. For in the first place, self-generation also exists on the specifically personal level as the maturing and constitution of the person and his or her character by free and moral action, by education, etc. The person does not generate himself by the mere biological processes common to the person and to plants and animals. Rather, the person generates and, in a certain sense, "creates" his own being by specifically personal actions and activities. Second, the diseases of the self-generating system of the living being have completely different, and devastating, effects on a human person than they have in non-personal organisms, for these effects are inseparable from rational human consciousness and cognition. Third, most self-generating aspects and healthy states of human life are inserted in contexts of human consciousness, without which their human and personal meaning would be incomprehensible; lack of an understanding of such contexts renders properly human and personal health unintelligible. This insertion of the biological sphere into human consciousness refers to the ability of persons to control with their freedom many aspects of this self-generation, but also to the recognition of those meanings and values which are associated with them and the response given to these. Fourth, the full health of the human self-generating aspects of biological life is not safeguarded when they merely function well biologically, but only when the meaning that they possess and should possess in persons is actually preserved, at least to a minimal degree.

Most dimensions of self-generation in humans are inserted into the personal dimensions of their lives and, therefore, most diseases of the self-generating functions of the human organism affect human consciousness. Human nutrition is not only regulated by unconscious processes, but also steered by conscious and free acts such as eating and free attitudes towards eating and reality at large. Illnesses therefore affect not only the purely physiological sides of nutrition but, for example, in *anorexia nervosa*, can also affect the ability to control food-consumption, the conscious and free attitudes toward eating, the reflective awareness of the effect of the consumption of food on one's appearance, etc. Similarly, abnormalities in the area of growth assume an entirely new meaning in the context of personal unhealth inasmuch as they may lead to complexes of inferiority, serious psychological problems and depressions, or even many social problems. And most of all, the sphere of procreation in persons, proceeding from conscious free relationships – related to the coming-to-be of persons and not of animals, subject to many moral or immoral forms of regulation – is in no way the same thing as in animals, and therefore can be affected by health or sickness in entirely new and mostly psychological and specifically personal forms.

Scientists and philosophers have identified other essential marks of life which allow us to understand new dimensions of health: for example, the interiority and ensouled character of the organism, its irritability, sensibility, contractibility, self-regulation, etc. Different philosophers have insisted on the response-structure of the organism and the fact that its reactions to the environment are never mechanical repetitions and never constitute a mere automatized, predetermined order, but rather constitute a striving to achieve one's own form. Thus the organism dissolves itself from the ties by which it is bound to its surroundings or nature, and shows an element of self-determination. With regard to all these dimensions of biological life, we can also identify the types of disturbances, reductions, losses of the corresponding abilities, etc. which constitute different forms of human un-health.

Other authors (Ehrenfels, Wertheimer, Conrad-Martius, Plessner) have added the feature of temporal *gestalt*-qualities and individual temporal rhythms according to which the growth and development of the organism take place. Regarding these, we can identify, as another part of health, the proper preservation and full actualization of these temporal patterns in the right rhythms, and can identify as disturbances of health the partial or total changes or losses of these patterns of development. It becomes apparent how both an empirical knowledge and some intuitive understanding of the "normal" temporal rhythms in which these *gestalt*-qualities ought to unfold is required in order to understand the specific nature of human health.

In the life of the organism, each part and organ possesses its function for the whole of the organism. Already in antiquity (for example, in Livius' report on Menenius Agrippa's fable and speech of the stomach and the limbs that ended the Plebeian secession) this feature of the organism was used in the imagery that compared states and other human communities to the organism. Also the Apostle Paul (1 *Cor*: 12, 12-27) uses the image of the organism, saying: "As the one organism possesses and needs many limbs and organs, the eye cannot say to the hand: I do not need you." This indicates not only some hierarchy but also some mutual dependency of all organs and parts of the organism on each other and of the whole on all of them.

We can speak of health and disease with respect to all of these properties of biological life, and it could be analyzed in detail how on the level of the person all these features of health and life take on a new meaning. For example, human health requires many uses of the hands in ways which the health of animals does not require, such as the ability to make pictures, to write, and to act in many other specifically human ways.

B. Entelechy as Teleological Directedness of Life and Health

Of the many other characteristics of biological life which, in their literal as well as in their analogous meanings, could become the starting point for comprehending the biological and specifically personalist dimensions of human health better, we choose here only one further feature of biological life and health, but a very central one. Spaemann and Löw (1981), Uexcüll (1921), Driesch (1928), and others have done much to rehabilitate the notion of teleology and the dominion of the question "why?" ("for the sake of which reason?") for the biosciences.

The attempt to identify the central mark of life in terms of finality combines finality with the self-containedness of the end in the organism, in an *entelechy*. This term – which comes from the Greek *en* (in), *telos* (end), and *echein* (to have) – is undoubtedly one of the most ingenious linguistic inventions of Aristotle (*De anima*), reflecting one of his most significant philosophical discoveries, and designates the "having within itself its own end" characteristic of *bios*-life.

The key notion of the influential General System Theory, namely the concept of an 'organized whole,' or 'organized totality,' also has much to do with *entelechy* in this sense although it rejects the very notion of *entelechy*.[15] One problem with this theory, which was shown by Conrad-Martius (1964) and Paulina Taboada (this volume) to amount to a subtle reductionism, is that it fails to account for the *inner* finality, and above all for the *dynamically self-forming* character of this organization.

The organism's dynamic teleological structure involves preserving its form through nourishment and oxygen-transfer,[16] but also recreating and transmitting it through regeneration and propagation. Conrad-Martius (1963) and Kant (1793) have admirably described the *mirandum* of this feature of biological life. Conrad-Martius compares this feature of life to what would be the case if an ingenious architect who produced extraordinary works of art succeeded in putting his idea into the colors, stones and other materials so that the work could paint and build itself.

Health, generally speaking, involves the intactness of all these moments. But again the need for a personalist concept of health emerges. In the first place, only a person understands the teleology of his own organic life and thus suffers on perceiving and experiencing its deviation from the pre-given teleological species-plan. Second, in the person the biological finality of the body-plan is in the service of a higher rational life which can use the limbs and members of the body, but can do so properly only when they are healthy. Third, there are entirely new senses of active self-generation on the level of personal life and hence personal health. For, in an analogous and even more profound sense, self-generation is not only characteristic of biological life, but is also a mark of the higher life of free beings who, through their value responses and free acts, in a certain way build themselves up and engender themselves in and through conscious acts.

C. The Essential Link Between Bios and Lived Body as a Clue to Understanding Human Health as Personal Health

Life as *bios* is inseparably linked to the physical sphere of some bodily beings.[17] Aristotle defines *bios*-life therefore as the act (or essential form) of an organic body (*De Anima* II, 412b). Quite independently from the definition of the terms, *bios*-life essentially demands from the kind of entity in question the relationship to a bodily being. This essential link between *bios*-life and body involves a tremendous variety of matter-life-relationships in plants and animals, culminating in the subtle life-processes characteristic of the human organism and such complex organs as the human brain or such stupendous systems as the human nervous system in its totality.[18]

Human biological life (*bios*), while it cannot be simply identified with the psychic and mental life of the human person, still interacts with it. The interpenetration of the *bios*-life and the mental personal life in the human body leads to the deep and intricate body-mind-union in human beings which involves a great variety of biological and simultaneously mental and psychic factors which are intimately connected with body-experiences, as they are intertwined with human sense perception, in the internal feeling and living of the body, in the expression of psychic life in the body, etc. (cf. Seifert, 1989).

In view of the inner link between *bios*-life and higher mental life, human health requires not only the intactness of all physiological conditions of these conscious and specifically human mind-body-relationships, but also the intactness of the specifically personal relationship to the lived body. Therefore any form of interruption or disturbance of this complex personal union of body and mind in man, whether for physical or psychic reasons, for example in the many phenomena of loss of body experience analyzed by Oliver Sacks (1995), is likewise a disturbance of health. In paralysis, or in the case of absence of the specifically human feeling of one's own body, different dimensions of this union are impeded or even lost. In quite another way, a disturbance of the relationship to the lived body gives rise to occurrences in which experiences that should provoke pleasure are accompanied by displeasure or nausea, as, for example, in the absence of appetite, aversions and allergies to the taste of good food, or in sexual disorders such as frigidity.

Few disorders on the human biological level leave the mind unaffected. And, in addition, mental illnesses and mental causes of psychosomatic disorders can affect the human body-experience and thus human health. Only in view of a properly personalist concept of health is it possible to account for the inner union of body and soul which pertains to human health, and to recognize all types of disorders which inhibit, reduce or destroy the organic unity of human bodily experience as the fulfillment or expression of personal acts.

V. HUMAN HEALTH AS THE INTEGRITY OF THE BODY/MIND RELATIONSHIPS

We have touched upon this important dimension of human health in various contexts and would overstep the boundaries of this paper were we to develop it here extensively. In light of the distinctions drawn elsewhere (Seifert, 1989) between three different types of body/mind relationships and their different modes, however, I would simply like to mention the crucial significance of these relationships for the concept of health. Human health involves the capacity to feel those experiences which can stem from bodily causes such as pain or pleasure, but it also requires a fundamental physical well-being which has a clear experiential dimension and is jeopardized by intense pain and the loss of basic well-being in its various dimensions. A more or less clear line allows us to distinguish here those dimensions of physiologically caused feelings of health that are indispensable parts of health because they belong to the basic physical and psychic well-being of a person from those that go beyond health as this basic physical well-being (such as intense physical pleasure). Also, the psychic states which can be caused by physiological events are partly *constitutive* of health, such as a fundamentally good mood as distinct from depressions, partly go *beyond* health, such as a superb good mood or ecstatic psychic *joie de vivre*, and partly constitute *unhealth*, such as serious psychic states of depression that are physiologically caused.

Physiologically caused destructions of mental capacities are certainly grave impairments of health. The same is true of certain causally produced irrational elements of intentional acts (Seifert, 1973).

Human health of the senses and of the mediating role of the body for our perception requires not only the chemical, electrical, and optical functioning of the body, but also the intactness of the specific role of the body in relationship to the receptive perception of the world. This aspect of health can suffer harm also in the form of 'hysterical blindness' and related phenomena analyzed by the young Freud. The same is true of other relationships "from the body to the mind," such as the ability to feel bodily pleasure and pain, to be "reached" and touched *as person* in the body, etc.

Also, the static body/mind relationships such as the body as physiological *condition* of knowledge, mental acts, physical actions, etc. have to be intact to preserve the health of the human person. Two of these static relationships (the body as expression of the human person and the body as potential expression of specific mental phenomena) we have already covered above, among the pre-biological aspects of health. But the body is also a static *condition* of human mental life, the *sine qua non*, without which man cannot think or decide, even though the body can never be the *cause* of knowledge or free action, as Socrates sees so clearly in the *Phaedo*. This "static relationship" has as well a specifically *human form* in the human person in whom the brain, for example, is not just the condition of certain animalistic operations, but of specifically personal intellectual and free acts.

Clearly, the third fundamental direction of the body/mind relationships, namely those that proceed from the mind and go toward the body, also need to be intact in their meaning and nature for the human person to be healthy. They involve not only causal psycho-physical influences of mental anguish or happiness on the physical well-being, aspects of health and disease which are explored by psychosomatic medicine, but also different relationships in which the person takes an initiative in speaking, acting, and the like. Here, the lived body is not only causally influenced by the mind but becomes a medium of conscious acts and actions. Just consider the role of the body in speech and conveying one's thoughts through language. Shakespeare (*Titus Andronicus*, Act III, Scene 1) expresses this admirably with respect to the tongue's function for speech when he has Lavinia's uncle lament her mutilation through the sons of the empress.

Thus it cannot be doubted that this third type of body/mind relationship, which comprises all relationships to the body in which the mind takes the initiative, is also of crucial significance for health. It involves all kinds of voluntary acting and speaking, as well as the expression of acts which originate in the psychic and spiritual life of the person but can incarnate themselves in bodily expressions. Of the body as *expression of mental life* we have already spoken, and may add here that this relation between mind and body does not only have its pre-biological dimension (which can also be found in the statue). Rather, inasmuch as the body is involved in intuitively manifesting the inner life of the person and in making it visible and audible in bodily expression, this dimension of health is intimately related to life and even to *specifically personal life*.

And it is at this point only that we reach the valid contributions of the action-theoretical models of health such as that of Nordenfelt. For indeed, the ability to carry out bodily actions of the most diverse sort is part of health, partly as those acts and actions themselves which we perform by means of the body are concerned (since an entirely passive person with the ability to act would still be sick), mostly, however, as far as the capacity and potentiality to act is concerned.

VI. MENTAL HEALTH AS THE CORE OF PERSONAL HEALTH

The previous considerations lead organically to the question of mental health and its link to the specific nature of personal being. Given the primary datum of health and mental health, we cannot really define the content of 'mental health' by reference to other things. Nevertheless, the nature of mental health can be characterized in the following ways:

1) Mental health is inseparable from the possession of rational consciousness and an (actually or potentially) awakened conscious state. If consciousness is unreachable for a person whose body-machine works, obviously this person is far from being "in good health."

2) Also, the intentional structure of human conscious acts (which aim consciously and meaningfully at their objects) provides a crucial key to comprehending a person's mental health. Conscious states without objects such as tiredness or euphoria are not enough to constitute mental health. Rather, an intentional meaningful relation to the world is a condition of a person's health.

3) Both on the level of non-intentional states and intentional rational relationships to the world, a certain amount of well-being, which includes a structural rationality explained below, and a certain degree of joy and *feeling well*, are constitutive of mental health (and, inasmuch as feeling well, also includes bodily feeling well, even of physical health).

Such a feeling of wellness of the subject objectively belongs to health, and has nothing to do with subjectivism. The aspect of consciousness and intentionality within *human* mental health certainly involves both the human *subject* and the consciousness of the subject, but it is an entirely real and objective factor. Therefore, one must reject the thesis that any subjectivist connotation flows necessarily from the WHO definition of health simply because it includes psychic (conscious) well-being in the notion of health.[19]

The objectivity of the notion of well-being that is part of health is obscured and denied only when the shift from objective well-being and that "feeling well" which objectively belongs to health to a merely subjective feeling of well-being is made. As soon as no objective criteria for the content and nature of feeling well are admitted, the definition becomes subjective. The subjectivist interpretation of health to which the WHO definition of health remains open, manifests itself today for example in the way in which the term "reproductive health services" is used so as to include the destruction of unborn life and the permanent destruction of the ability to conceive children, when the mother feels subjectively good about having abortion or sterilization performed. Here what objectively constitutes a destruction of health or life is called health – based on an entirely relativistic notion of the contents of well-being and "feeling well" which constitute health.

4) Mental health can also be described as the uninhibited flourishing of, and simultaneously as the potentiality toward, the entire sphere of mental and rational life of finite and embodied (human) persons. Mental health allows not only the performance of many intentional acts, for example abstraction, but also the intactness of all fundamental faculties of the human person and the ability to exercise them: intellect, will, instincts, feelings, the execution of spiritual acts, including a conscious living of one's own body as one's own in a personal way, unimpaired perception, cognition, and emotions. This dimension of mental health is the opposite of various forms of mental illness such as schizophrenia and mental retardation.

Mental health as the nucleus of personal health also includes the element of having at one's free disposal the different perceptual, imaginary, intellectual, volitional and affective powers of personal life, as opposed to having one's consciousness invaded by strange images, hallucinations, irrational ideas, or irresistible compulsions. As the person is characterized also as a center of rational-intentional life, the psychic health of human affective life is a necessary part of the health of human persons *qua* persons. This dimension of the person's health stands in contrast also to the crippling of the emotional life by the inability to feel, affective frigidity and other emotional disorders (cf. Roa, this volume).

When mental illness takes the form of affective disorders and depressions, it interferes with the person's possession (or having at his or her disposal in a rational way) of his or her emotional life. Of course, it belongs to the essence of perfectly normal affective experiences that they are not in our free control. Yet in the case of severe affective disorders, we lose a type of control of reason over the affections which is part of the person's health.

5) Given the character of the human person as a *zoon politikon*, as a social being, mental health also involves having the ability to enter interpersonal relations. Mental illness is correlatively characterized as abnormal behavior, actions, feelings, or imaginations which disturb the relations of mentally ill persons with others and with society, and which interferes with their ability to work and love (cf. Kazdin, 1996a, 1996b).

6) Referring back to the idea of the fulfillment of the species-plan as a condition of health, we can add: Mental health presupposes a conscious life in accordance with the essence and species-plan of the human person, or rather with the objective and self-transcending structure of the human person. To hold this presupposes a notion of the objective essence of the person and a rejection of Sartre's idea that because the human person is not simply lived by his nature and his nature is not totally fixed but profoundly open to the influence of freedom, the human person possesses no nature at all but creates his own essence (cf. Seifert, 1987, 1996a). Instead, if the human person, the goals and values he ought to realize, and the essence of his various acts such as judging, conceiving, perceiving, loving, etc. each have their immutable essential form, then this species-plan is the very foundation of health.

7) Mental health cannot be divorced from the dynamic unfolding of human nature in time. Mental health therefore involves a certain actualization of conscious and personal life suitable for an individual's

stage of development, and even more the potential to live this life without crippling obstacles.

8) Mental health also involves the possession of a fundamental self-consciousness, the acceptance of the 'I' (Roa, in this volume; Jahoda, 1958), as well as a balanced assessment of the world and other persons.

Various mental disorders, such as psychogenic amnesia, pathological self-hatred and self-contempt, or guilt-complexes and complexes of inferiority can interfere with this dimension of mental health.[20]

9) Mental health likewise involves a certain ability to retain the identity of one's personality, convictions, ways of feeling, etc., in different situations and environments, and at the same time an openness to change where objective factors require it. Those forms of mental illness which one calls "adjustment disorders" can include, on one extreme, a total loss of self-identity and a high degree of adaptation to every social wind around oneself. In its radical form, this mental illness constitutes a certain loss of self-identity. On the other extreme, we find mental illness in the form of a complete withdrawnness from the natural influences and forms of legitimate adaptation to one's environment, such as in total autism or organic disorders which lead to a total incapacity to be influenced by one's surroundings.

10) Mental health includes additionally the possession of conscious, intentional goals, as well as some equilibrium between them and the fundamental capabilities of reaching these goals. Nordenfelt (1986) adds, in a modification of the equilibrium theory of health developed by Whitbeck (1981) and others, that the existence of certain goals, at least minimal ones, is also required for health. In itself, however, this is insufficient to account for health. For Nordenfelt's idea of a "minimal set of goals" as an additional factor that is irreducible to equilibrium cannot be understood merely quantitatively. It also must take into consideration the *quality* of these goals, especially their objective rootedness in the nature of the subject and their adequacy to reality.

11) Mental health also consists in the ability to perceive and understand things properly as they are, to relate to objects adequately, and to distinguish good and evil. Mental health embraces a certain fundamental structural rationality of one's intellectual, volitional, and emotional life and acts. This rationality is not yet the higher rationality found in adequate knowledge of reality or moral goodness – which includes much more than mental health – but is rather a "basic rationality" which prevents one from holding totally absurd errors, failing

to make the most elementary distinctions, willing impossible or completely neutral things with great intensity, or feeling deep pain where the object in no way calls for such a response. Without understanding the nature and rationality of due-relation and value-response (cf. Hildebrand, 1978), this dimension of rationality of intentional acts can hardly be conceived. This rationality excludes, for example, the fear of entirely harmless things, strong hatred of completely innocuous animals, or intense anger toward wholly innocent persons.

Beyond consciousness and intentionality as such (which characterize also sick fantasies), a fundamental conformity or adequacy of human acts to their objects is a prerequisite for human mental health. A human person who would be incapable of relating adequately to the objects of his intentional acts, would not fear earthquakes but doorknobs, would never feel love but only disgust and hatred for persons worthy of love, but intensely love snakes and monsters, would lack mental health and indeed be mentally sick.

There are two types of adequacy of intentional acts to their objects, however: one is the criterion of mental sanity, and the other is the adequacy of knowledge, or of the truth of love or virtue. We could in a certain way define mental health in these terms: A person's mental health involves some fundamental awareness of the real nature of the intentional objects of one's acts and some fundamental meaning in the intentional relation to them that is common to normal good and evil acts of the will, and both to true and erroneous but "normal" ideas about things.

Mental health also involves the ability to make fundamental distinctions, especially of the real from the unreal and merely imagined, of other ontological modalities (such as the possible and impossible, probable and improbable), and of good from evil. This conformity to reality contained in mental health is not so much the opposite of mental retardation as of certain kinds of mental illness such as paranoia. Similar results can also flow from other psychotic disorders, the most prominent type of which is schizophrenia, in which severe disorders interfere with a person's thinking, distort his perceptions and emotions, and reduce the ability to make the aforesaid distinctions.

VII. PERSONAL HEALTH AND GOODS HIGHER THAN HEALTH: ON THE LIMITS OF HEALTH AND ITS DEPENDENCE ON WHAT LIES BEYOND IT

A drug addict is not in possession of the free control of his actions. As a victim of his addiction, he becomes unable to foster family life and love; he is sick. Certain dimensions of mental health are the fruit or even a dimension of moral goodness, just as certain aspects of mental sickness, far from rendering freely perpetrated moral evil impossible, are rather the direct consequences and accompanying aspects of freely perpetrated immoral acts or attitudes.

There is an even higher, analogous and archetypical sense of health, in which moral goodness itself may be called health. Plato uses the term in this sense when he speaks of the souls of evil persons. They will have to appear after death before the throne of Radamanthys, the Judge of souls, who will find of many souls including those of "the Great King or of other kings and princes" that "nothing healthy is in them" (Plato, *Gorgias*, 524 e). Here health is equated with moral goodness, and disease with moral evil, which makes good sense if we consider that the deepest essence of health is the flourishing and actualization of life, and that a person's life in the deepest sense is inseparable from moral goodness. From this it follows that health in the deepest sense is inseparable from moral goodness, for only in the morally good person can life flourish in its entirety. The mental harmony of the soul of a sadist can only be restored by a moral conversion, as Vitz showed in his paper on "Hatred and Forgiveness in Psychotherapy" (1996).

Mental health in the ordinary sense, while it envelops many relationships to the moral sphere besides that of being a condition for moral acts, is yet distinct from moral goodness itself. Notwithstanding this distinctness, there is an inner logic which leads the excessively proud man to become a megalomaniac, or a certain type of hate-filled person to fall into the pathological "splitting" described by Vitz (1996), in which he sees himself as all-good and his hated opponent as all-evil.

All great poets knew that terrible pride and unresolved feelings of guilt, etc. can have truly maddening effects (Shakespeare's Lady Macbeth being an example), and that they can destroy even the noblest minds in such a way that they become truly sick in the clinical sense. Think also of the well-explored 'post-abortion-syndrome' in which many neurotic symptoms proceed clearly from unresolved feelings of guilt.

There are moral evils which are also spiritual and mental diseases. In some cases morally good acts and attitudes are themselves part of mental health and morally evil ones are part of mental illness. Imagine a man or a woman who chooses to doubt every piece of knowledge that is not absolutely certain and indubitable and who therefore does not take one step in trust. Such a person will utterly distrust any other human person because he or she can never know with indubitable certainty the secret intentions and thoughts of others. If such an attitude is pushed to its extreme, it makes any social and inter-human relations impossible.

The person who refuses any fundamental trust toward fellow-human beings and a human person who distrusts completely the author of all being violates the fundamental trust man owes to fellow-humans and above all to God and His providence. The refusal of a morally good trust will, however, also lead to a totally neurotic and indeed psychotic condition in which a person is wholly cut off from all human relationships and tormented by fears, suspicions, and finally succumb to an almost total paralysis of his actions. A total misanthrope distrusting other human persons down to the innermost thoughts of his heart could not even accept the food from his grocery store or from his family members and would die of starvation. His social relations would become practically impossible and crippled in an unbearable manner. It would be hard to exaggerate the state of mental unhealth that would result from such an attitude.

Mental health also has certain metaphysical presuppositions. For if God were out to destroy us, a state of total distrust, which would give rise to an insuperable anguish and depression, would be an adequate response to the ultimate reality. At the same time, this "logical" state of mind would constitute supreme mental unhealth. We could say with Nietzsche that no man ever had the strength to live in such an atheistic world, drawing all logical consequences from his position existentially.

In like fashion, if all other human beings were liars, deceivers, and murderers, a total distrust in humans, which would lead to mental unhealth and madness, would be adequate to reality. Therefore, a healthy mental life also makes certain assumptions about some goodness or kindness in other human persons.

Gratitude, humility, and the readiness to forgive offenses are likewise conducive to mental health, although this cannot be shown in detail here. Many other morally good qualities are not only conditions but even a necessary part of mental health, and the related moral evils, when lived existentially and with all consequences, constitute states of mental illness.[21]

Nevertheless, in spite of the interwovenness of personal health with the entirety of the objective goods for persons, health must be clearly demarcated from these higher goods and never confused with them. To counter the utopian spirit of the WHO definition of health, we must not define health in terms of "the complete physical, mental, and social well-being" of the human person nor identify it with those physical, psychic, social, and spiritual goods of persons which lie beyond the sphere of health, even health taken in its wider sense. The distinctness of the category of health gets lost and the concept corrupted when all sources of human, social, and religious well-being are defined as "parts of health." Correspondingly, the objective nature of human life and personhood forbid viewing all suffering, aesthetic disvalue, pain, and deficiency of social relations as constituting or causing a lack of health.

First of all, not only sickness and injury cause pain and suffering. With regard to physical pain, for example, pregnancy and childbirth, although causes of physical discomfort and pain, are not diseases and do not involve unhealth (although the accompanying physical discomfort itself, just as that resulting from old age, involves some absence of the fullness of health).

Second, not all physical pain and even less all mental suffering constitutes unhealth or is a sign thereof. On the contrary, a complete insensitivity to physical pain is normally even a sign of some disease. Some mental suffering, for example that experienced in grief, is even a sign of health, as we can see after our analysis of adequacy to goods and evils as a dimension of mental and personal health. Suffering that results from an adequate value response to evil, and even physical pain which accompanies normal states of life (such as old age) or activities that are in perfect harmony with human nature (such as childbirth) do not constitute opposites to health. This does not preclude that, under certain aspects, *all* pain and suffering constitute obstacles to a state of *perfect* health (taken in the wider sense distinguished above), and that therefore an *extreme* state of health would also include the absence of all pain and suffering. Three distinct factors are responsible for the delineation of those pains and deformities which constitute contrasts to health from those which do not: a) the question of whether the *basic* form of the human person feeling bad or well is absent; b) the question of whether the respective deformities are against human nature or the human species-plan; c) the question of the adequacy or inadequacy of some pain and suffering over deformities which do not endanger health.

Third, all the higher forms of the functioning of the body and senses and the aesthetic beauty of appearance, which exceed a sheer minimum of normality, lie beyond health.

Fourth, knowledge, science, and the goods which result from them lie beyond health, while they belong to a "complete mental well-being."

Fifth, moral goodness is specifically more than health and resists any attempt to be subsumed under the notion of health, especially when the latter is conceived of primarily as a vital value, as was the case in Nazi-ideology and its extolling of health over and against deeper personal values.

Sixth, the health of an individual must neither be conceived of as including all interpersonal relationships and the happiness resulting from them, nor be dissolved into an infinite series of relationships to the environment, society, etc.

Seventh, and more generally speaking, many human goods (such as exuberant vitality, friendship, love, the life of those we love, knowledge, wisdom, virtue, exceeding beauty) clearly surpass the realm of health and are goods of an entirely different nature. To subsume them under health evidently neither does justice to them nor to health.

Therefore, health does not encompass all bodily beauty and pleasure, all sources of human happiness in knowledge, love, and beauty, nor are all moral evils, deformities and pains opposites to health. We have to criticize any such utopian and unrealistic concept of health as an all-encompassing good of the human person.

In spite of its distinction from higher goods, especially moral goods, health has a close relationship to them. Reminding ourselves of what Socrates says in Plato's *Apology* about wealth and fame, we may say analogously that health is not the highest good for man, and yet "from virtue [...] wealth and all other goods come" (30b), including health. This is even much more true for health than for wealth, for a person not given to the quest for truth and the good cannot be entirely healthy as a person. A demonically evil person will also be profoundly mentally sick. The health of a person ultimately presupposes a full givenness to goods which transcend health, i.e., only the person who seeks the True, the Beautiful, and the Good in all their forms and especially in their highest forms, can be fully happy and healthy *qua* person.

VIII. CONCLUDING REMARKS

This paper sought to establish the threefold relationship of human health to pre-biological elements, biological elements and the specifically personal dimensions of the human being which are constituted most specifically by mental health. Moreover, I tried to show that all the pre-biological elements of health are parts of specifically *human* health only by their relationship both to biological life and to specifically personal life, and that all biological dimensions of human health are parts of specifically human health only in virtue of their insertion and integration into the sphere of personal being and mental acts and the specifically human significance of the body. On the other hand, while including many aspects of conscious life and aesthetic good in the broader concept of health used in this paper, I insisted on objective nature and species-plan as foundations not only for the biological but also for the personalist concept of health. All of them are parts of health not in terms of a merely statistically conceived species-plan, but in reference to an objective and intelligible nature of the human being, which has both empirically discovered and *a priori* known (objectively necessary) sides. Finally, while health, even in the broader sense of the term used in this paper, includes many aspects of consciousness and social relationships or the capacity to have them, it is not the all-encompassing good of man, while at the same time it cannot be realized without some realization of those goods, for example moral goodness, which lie beyond health.

International Academy of Philosophy
Principality of Liechtenstein

NOTES

[1] This does not exclude the possibility of certain vague analogies to personal health in the psychic health of animals or even in the environmental aspects of the health of plants.

[2] See also the critique of these views in Nordenfelt (1986), pp. 282 ff.

[3] "German Measles," Microsoft (R) Encarta (1995).

[4] See the article "Plastic Surgery," Microsoft (R) Encarta (1995).

[5] Boorse's (1975, p. 559) idea of the species-plan as a condition of understanding health is quite applicable here, even if the reduction of this species-plan to a mere statistical value is wholly to be excluded.

[6] On the entirely irreducible essence of expression in this sense, see Max Scheler (1966) and Dietrich von Hildebrand (1977).

[7]   I am not aware of any other author having introduced this distinction between pre-biological and biological aspects of health.

[8]   "Modern biology ... can only attempt to approach the living by simulating it. Its mistake consists in believing that the simulation is the original" (Spaemann, 1994, p. 86; translation mine).

[9]   On this notion of "essential core" – introduced by Jean Hering and Roman Ingarden – which signifies the inner principle of unity of an essence see Josef Seifert, 1977, and 1996b, particularly ch. 1.

[10] Serani (1986) appears not to admit of organic life independently of nutrition. Also in cryo-conservation health may be preserved. See on this Kenneth B. Storey and Janet M. Storey (1990).

[11] Also the form of self-motion in the animal requires entirely new categories in comparison with those of plants. See Conrad-Martius (1963). See also my comments on Conrad-Martius' fascinating analyses of these phenomena in Seifert, 1973.

[12] A human person is never reducible to a merely vegetative life. The permanent vegetative state should therefore be called a "permanent state of deactualization of higher activations of human life." Cf. Josef Seifert (1989).

[13] "ut...libertate utentur, cuius proprium est sic vivere, ut velis...."

[14] Think also of other champions of freedom in ancient philosophy, such as Carneades or Epicurus. With respect to their clear recognition of freedom, Bonaventure, Descartes, or Sartre were hardly more outspoken than thinkers such as Augustine.

[15] See Purola (1972). Other names of contemporary defenders of some version of General System Theory today include Noack, Anderson, and Kaplun.

[16] For this reason, Serani (1986) prefers to speak of morphogenesis rather than of nutrition and of growth when pointing at the essential characteristics of organisms.

[17] Aristotle's words *(De Anima,* 414 a 19-20) apply here: ."..the soul cannot be without a body, while it cannot *be* a body." (Translation mine.)

[18] One must distinguish sharply the plant's material aspects from the animal body, and both from the human body of which Hedwig Conrad-Martius (1963) rightly says that it alone can properly be called *"Leib"* in German.

[19] "Health is a state of complete physical, mental and social well-being and not merely the absence of disease or infirmity" (from the Preamble of the *Constitution of the World Health Organization,* adopted by the International Health Conference in New York, 1946).

[20] On the many meanings of the term "inferiority complex" and "feelings of inferiority" see Seifert (1995a).

[21] See on this my (1996a), where I distinguish more sharply entirely different relationships between morality and mental health.

BIBLIOGRAPHY

Adler, A.: 1973, *Der Sinn des Lebens,* Fischer Taschenbuch Verlag, Frankfurt a. M.

Aristotle: 1971, *De Anima (On the Soul),* J. A. Smith (trans.), in R. McKeon (ed.), *The Basic Works of Aristotle,* Random House, New York, pp. 533-603.

Aristotle: 1961, *Eudemian Ethics,* H. Rackham (trans.), Loeb Classical Library, Harvard University Press, Cambridge, Massachusetts, pp. 190-477.

Bernard, C.: 1966, '*Leçons sur les phénomènes de la vie communs aux animaux et aux végétaux*,' J. Vrin, Paris, photo reproduction of the first edition, J.-B. Baille, Paris, 1879-1885.

Bertalanffy, L.v.: 1952, *Problems of Life. An Evaluation of Modern Biological and Scientific Thought*, Watts and Co. Ltd., New York.

Bonaventura, *Doctoris Seraphici S. Bonaventurae Opera omnia*, edita studio et cura PP. Collegii a S. Bonaventura, ad Claras Aquas (Quarracchi) ex Typographia Collegii S. Bonaventura, 10 volumina (1882-1902).

Boorse, C.: 1975, 'On the distinction between disease and illness,' *Philosophy & Public Affairs* 5, 61.

—— 1977, 'Health as a theoretical concept,' *Philosophy of Science* 46, 559.

Caplan, A.L. et al.: 1981, *Concepts of Health and Disease. Interdisciplinary Perspectives*, Addison-Wesley Publishing Company, London/Reading, Mass.

Compton's Interactive Encyclopedia. Copyright (c) 1996, Compton's NewMedia, Inc.

Conrad-Martius, H.: 1963, 'Die Seele der Pflanze,' in H. Conrad-Martius (ed.), *Schriften zur Philosophie*, vo. 1., Eberhard Avé-Lallement, Kösel, Munich, pp. 276-362.

—— 1964: 'Präformismus in der Natur,' in H. Conrad-Martius (ed.), *Schriften zur Philosophie*, vol. 2, Eberhard Avé-Lallement, Kösel, Munich, pp. 153-173.

DeBakey, M.E.: 1995, 'Surgery,' Microsoft (R) Encarta 1995.

Driesch, H.: 1928, *Philosophie des Organischen* (Gifford-Lectures), Verlag von Quelle & Meyer, Leipzig.

Engel, G.L.: 1977, 'The need for a new medical model: A challenge for biomedicine,' *Science* 196.

Engelhardt, H.T., Jr.: 1981, 'The concepts of health and disease,' in A.L. Caplan, H.T. Engelhardt, and J.J. Mc Cartney, *Concepts of Health and Disease. Interdisciplinary Perspectives*, Addison-Wesley Publishing Company, London/Reading, Mass., pp. 31-46.

Frederick A. M.: 1995, 'Spinal column,' Microsoft (R) *Encarta*.

Heidegger, M.: 1927, *Sein und Zeit*, Niemeyer, Tübingen.

Hildebrand, D.v.: 1960, *What is Philosophy?*, 2nd ed., Bruce, Milwaukee.

—— 1977, *Ästhetik*, 1. Teil, Gesammelte Werke, vol. V, Kohlhammer, Stuttgart.

—— 1978, *Ethics*, Franciscan Herald Press, Chicago.

Jahoda, M.: 1958, *Current Concepts of Positive Mental Health*, Basic Books, New York.

Kant, I.: 1793, *Kritik der Urteilskraft (Critique of Judgment)*, F. T. Lagarde, Berlin.

—— 1968, *Träume eines Geistersehers (1777)*, in *Kants Werke, Vorkritische Schriften II, 1757-1777*, Vol. IV, Akademie-Textausgabe, Walter de Gruyter & Co., Berlin.

Kass, L.: 1981, 'Regarding the end of medicine and the pursuit of health,' in A. Caplan et al., *Concepts of Health and Disease. Interdisciplinary Perspectives*, Addison-Wesley Publishing Company, London/Reading, Mass., pp. 3-30.

Kazdin, A. E.: 1996, 'Mental illness,' *Compton's Interactive Encyclopedia*.

—— 1996a, 'Mental retardation,' in: *Compton's Interactive Encyclopedia* 1996.

Lederer, F. L.: 1995, 'Ear,' Microsoft (R) *Encarta*.

Louis J. V.: 1995, 'Orthopedics,' Microsoft (R) *Encarta*

Nordenfelt, L. and Lindhal, I.(eds.): 1984, *Health, Disease and Causal Explanations in Medicine*, D. Reidel Publishing Company, Dordrecht.

Nordenfelt, L.: 1986, 'Health and disease: Two philosophical perspectives,' *Journal of Epidemiology and Community Health*, 41, 281-284.

—— 1987, *On the Nature of Health*, D. Reidel Publishing Company, Dordrecht .

Pjörn, I.: 1984, 'An equilibrium theory of health,' in L. Nordenfelt, I. Lindhal (eds.), *Health, Disease and Causal Explanations in Medicine*, D. Reidel Publishing Company, Dordrecht.

Plato: 1961, *Collected Dialogues*, E. Hamilton and H. Cairns (eds.), Princeton University Press, Princeton, N. J.

Purola, T.: 1972, 'A systems approach to health and health policy,' *Medical Care* 10, 5.

Roche, J.: 1995, 'Fetus,' Microsoft (R) *Encarta*.

Sacks, O.: 1985, *The Man Who Mistook His Wife for a Hat and Other Clinical Tales*, Summit Books, New York.

Scheler, M.: 1966, *Der Formalismus in der Ethik und die materiale Wertethik*, Francke, Bern and Munich.

Seifert, J.: 1973, *Leib und Seele. Ein Beitrag zur philosophischen Anthropologie*, A. Pustet, Salzburg.

—— 1977, 'Essence and existence. A new foundation of classical metaphysics on the basis of "Phenomenological Realism," and a critical investigation of "Existentialist Thomism",' *Aletheia* I,1, pp. 17-157; I,2, pp. 371-459.

—— 1987, *Back to Things in Themselves. A Phenomenological Foundation for Classical Realism*, Routledge, London.

—— 1989, *Das Leib-Seele Problem und die gegenwärtige philosophische Diskussion. Eine kritisch-systematische Analyse*, Wissenschaftliche Buchgesellschaft, Darmstadt.

—— 1990, 'Beauty of higher forms (second potency) in art and nature,' in *Du Vrai, du Beau, du Bien. Etudes philosophiques présentées à Evanghélos Moutsopoulos*, Librairie Philosophique J. Vrin, Paris, pp. 171-183.

—— 1991, 'Objektivismus in der Wissenschaft und Grundlagen philosophischer Rationalität. Kritische Überlegungen zu Karl Poppers Wissenschafts-, Erkenntnis- und Wahrheitstheorie,' in: N. Leser, J. Seifert, K. Plitzner (eds.), *Die Gedankenwelt Sir Karl Poppers: Kritischer Rationalismus im Dialog*, Universitätsverlag C. Winter, Heidelberg, pp. 31-74; and *ibid.*, 'Diskussion,' pp. 75-82.

—— 1992, 'The objectivity of beauty in music and a critique of aesthetic subjectivism,' *XRONIKA AISQUETIKHS, Annales d'Esthéthique*, Tom. 31-32, 33-61.

—— 1995a, 'Inferiority complex and response: Alfred Adler – Discoveries and errors,' in James M. DuBois, *The Nature and Tasks of a Personalist Psychology*, University Press of America, Lanham/New York/London, pp. 87-110.

—— 1995b, 'El hombre como persona en el cuerpo,' in *Espiritu* 54, 129-156.

—— 1996a, 'Morality and mental health,' in James DuBois (ed.), *Moral Issues in Psychology*, University Press of America, Lanham/New York/London.

—— 1996b, *Sein und Wesen*, Universitätsverlag C. Winter, Heidelberg.

Serani, A.: 1986, *L'Etre Vivant selon la perspective réaliste. Propos pour les fondements d'une biophilosophie*, Université de Toulouse le Mirail U.E.R. d'études philosophiques et politiques, Toulouse.

Shakespeare: *Titus Andronicus* (Complete Works on CD-Rom, HP).

Spaemann, R.: 1994, 'Zum Begriff des Lebens,' in G. Kockott, H. J. Müller (eds.), *Sichtweisen der Psychiatrie*, Zuckerschwertverlag, Munich 1994, pp. 84-89.

Spaemann, R. and Löw, R.: 1981, *Die Frage Wozu? Geschichte und Wiederentdeckung des teleologischen Denkens*, Bertelsmann, Munich.

Storey, K. B.& Janet M.: 1990, 'Frozen and alive,' in *Scientific America* (Dec.), pp. 62-67.

Uexcüll, J. v.: 1921, *Umwelt und Innenwelt der Tiere*, Verlag 5. Julius Springer, Berlin.

Vitz, P.: 1996, 'Hatred and forgivenness in psychotherapy,' in James DuBois (ed.), *Moral Issues in Psychology*, University Press of America, Lanham/New York/London.

Whitbeck, C. A.: 1981, 'A theory of health,' in A.L. Caplan, et al. *Concepts of Health and Disease. Interdisciplinary Perspectives*, Addison-Wesley Publishing Company, London/Reading, Massachusetts.

Wojtyìa, K.: 1979, *The Acting Person*, Reidel, Boston.

World Health Organisation (WHO): 1946, Preamble of the *Constitution of the World Health Organization*, adopted by the International Health Conference in New York.

SECTION TWO

HEALTH AND HUMAN WELL-BEING

H. TRISTRAM ENGELHARDT, JR.

HEALTH, DISEASE, AND PERSONS:
WELL-BEING IN A POST-MODERN WORLD

I. INTRODUCTION: HUMAN HEALTHS AND WELL-BEINGS

There is no generic human well-being. All actual human well-being has content and is specific. This is the case because well-being is contextual and there are numerous contexts within which human well-being can be defined. This is also the case because well-being must be specified in terms of a particular content-full understanding of values or human flourishing. To talk concretely of well-being is to specify a context and to accent some values over others, to endorse certain notions of appropriate human projects over others. Because there are divergent contexts within which to understand human well-being and because there are diverse accounts, understandings, and rankings of values and of appropriate human projects, a scrutiny of notions of health and disease reveals that they are in the plural.[1] There are numerous competing accounts of the good life and of human flourishing. This plurality is reflected in the numerous understandings and construals of health and disease.[2] The role of culture-dependent values in these differences is accented when particular understandings of disease and health are incorporated with health-care systems that must inevitably embody decisions as to what states of affairs should be the focus of appropriate medical concern and medical treatment.[3] The framing of an all-encompassing societal health care system inevitably discloses tensions among competing accounts of health and disease.

A societal account of health runs aground on the divergence among persons and communities regarding what it is to flourish as a human and as a person. This divergence is at a minimum to be understood in terms of an epistemic failure: it is not possible by sound rational argument to establish the canonical, content-full account of the right or of the values that should frame a vision of human flourishing. In addition, physical well-being, not to mention mental well-being, is contextual and goal-dependent. To realize physical well-being in a particular context, environment, or niche is to be well adapted to that niche in the sense of being able to realize the goals at stake in that environment. To give an

P. Taboada, K. Fedoryka Cuddeback and P. Donohue-White (eds.), Person, Society and Value: Towards a Personalist Concept of Health, 147–163.
© 2002 Kluwer Academic Publishers. Printed in Great Britain.

account of physical well-being or adaptation, one must specify the environment and specify the goals. It is one thing to be well-adapted to hunt in a tropical environmental niche, and another thing to be well-adapted to hunt in the Arctic. It is another thing to be well-adapted to read books in either of those environments.

This is not to say that one cannot frame a value-unspecified account of human well-being or health as a way of identifying the various and divergent concepts of human well-being and health. For example, one can define health as optimal adaptation, leaving unspecified the context of adaptation or the goals of adaptation. In this fashion, it would make sense to speak of a concept of health or well-being while at the same time holding that the concept is strategically ambiguous until one endorses a particular account, moral sense, or understanding of values. In general, one can identify a structure in terms of which to recognize well-being or health, while at the same time recognizing that there is no concept that can provide moral, value, or policy guidance or direction. If there is a genus of well-being or health that includes the numerous understandings of well-being and health, then it is an empty function that does not require or presuppose any particular hierarchy or ranking of values or goods.

One must specify the goals of adaptation in order to give content to well-being. The goods at stake specify what it is to be well. When considering biological adaptation, the goal at stake is usually inclusive fitness: the ability to effect the transmission of one's genes to the next generation, even if one does not oneself have offspring. Success is reproductive success, a form of success that does not possess a self-evident claim on our energies. After all, in the long run all species die out. Biological history is a series of disasters and lost species. Why should one have an overriding commitment to biological adaptation in the sense of reproductive success? For example, even if one were to determine that intellectuals are less likely to be successful reproducers than non-intellectuals, one might still wish to support the intellectual life, despite its encumbering human survival. The problem is why any particular set of goods or account of human flourishing should be taken as overriding.

If it is not possible to turn either outward to nature or inward to the character of humans and by secular philosophical means to discover the appropriate account of human well-being, then the pursuit of an understanding of human well-being will be the result of explicit and implicit agreements among persons. To realize a social endeavor such as medicine, agreements will need to produce various collaborating groups who direct their energies towards the pursuit of particular understandings of human well-being. If understandings of human flourishing diverge, and if persons are free to collaborate, they will frame different communities of agreement and understanding regarding how to achieve human well-being. It is not just that, if one examines the character of medical knowledge, one will discover a plurality of understandings of well-being. In addition, this plurality of understandings discloses the possibility of a plurality of social understandings of the goals of medicine and the realization of human well-being. Again, if persons are left to collaborate freely, there is no reason to suppose that one will not be confronted with numerous alternate social constructions of medical reality.

In our contemporary world, this invites the emergence of parallel health care delivery systems on the model of the parallel spiritual care delivery systems one finds in the Federal Republic of Germany. If one considers the analogy, however distant, between the Evangelical and Roman Catholic approaches to providing state-supported spiritual welfare and various approaches to the provision of health care, one can envisage the possibility of the parallel pursuit of alternate understandings of human well-being, as understood within the context of medicine. Notions of healths would then not be societal, but communal. One would be confronted with numerous parallel and competing projects of realizing health and well-being. The situation is post-modern in the sense of being framed in the absence of a universal narrative or account of health and well-being, and instead with the guidance of numerous, divergent, and at times incommensurable narratives or accounts among which one cannot choose in a principled fashion on the basis of a general secular sound rational argument.

This essay begins by showing the eradicably plural character of understandings of human well-being and flourishing. This involves exploring the value-laden character of concepts of human well-being, health, and disease to disclose why there cannot be a single canonical secular account. This argument turns on a form of epistemological but not metaphysical skepticism. The argument does not deny that there is a superior or appropriate account of human well-being, only that it can be established as governing by an appeal to sound rational argument. This essay then draws out some of the consequences of this state of affairs for the framing of parallel health care systems as parallel medical social realities. In the absence of a single, canonical, secular understanding of human well-being, disease, and illness, both individuals and communities should be at liberty to realize their own accounts, to give them substance and reality through alternative health insurance schemes, reflecting different understandings of human well-being.

II. HUMAN WELL-BEING IN A POST-DARWINIAN WORLD

Against a pre-Darwinian view of man's Divine design, it might have been presupposed that the character of human well-being and therefore the character of human illness and disease can be discovered. If human nature is regarded as having a design or being directed toward certain goals that human persons should pursue, then a canonical, content-full account of human health and well-being could be disclosed. Lack of agreement in this area suggests that this design and these goals are not disclosable in general secular terms. Indeed, contemporary reflections regarding the nature of well-being, illness, and disease reveal different and competing accounts. There is no concurrence about the bases for judging states of affairs to be states of disease or illness because there is no concurrence concerning how one might understand the appropriate design or goals of humans. This is as one would expect. To recognize a state of affairs as one of well-being or health, or alternatively as one of disease or illness, requires bringing to our appreciation of the human condition a particular normative account. Again, one need only consider what is involved in determining whether humans are healthy in the sense of well-adapted. Any such determination requires specifying the environment to which they are adapted and the goals to be achieved. We precisely fail to agree regarding what particular content-full goals ought to be guiding in determining successful human adaptation, as well as in identifying the canonical reference environment.

The range of understandings of appropriate human adaptation and excellence can be appreciated by considering germline genetic engineering and the prospect of being able to restructure human biological and psychological nature. The possibility of reengineering human nature serves as a heuristic for identifying the lack of agreement regarding the nature of human flourishing. Because in considering germline genetic engineering one must take the perspective of a designer of human nature, this heuristic underscores the presence of numerous alternative accounts of how one might refashion the human genome in the pursuit of human well-being. A choice about how to proceed in redesigning human nature requires accepting particular goals. Confronting such a choice forces one to take account of the diversity of understandings of human flourishing.

The first set of challenges to establishing a single account of human flourishing to guide human genetic reengineering is related to a bioengineering design problem. One likely cannot fashion an organism that can do all that everyone would want. Instead, it is very probable that one can produce an organism that is better adapted to some, but not equally adapted to all environments. The more one envisages a human organism adapted to all environments, the more one envisages a god, rather than a finite being with particular abilities for particular tasks. If one recognizes that an organism will be able to do better in some environments rather than others, one is forced to choose among environments for adaptation in designing humans. For example, should humans be designed to be ideal outdoorsmen able to survive well in most terrestrial climates without carrying considerable backpack equipment, etc., versus being designed with an ability to drink substantial measures of good bourbon, smoke good cigars, chew good tobacco, and live a primarily sedentary life in closed rooms with few, if any, adverse health consequences? The point is that, depending on the primary environment and lifestyle to which one thinks humans ought to be adapted, one will fashion one sort of human organism rather than another.

Here one returns to a fundamental challenge: specifying what it is to be a well-adapted human organism. Much has to be filled in to give sufficient content. One is confronted not simply with a choice between lifestyles (i.e., between a sedentary and active lifestyle), but with specifying the kind of environment within which one wishes humans easily to flourish. For example, should such a reference environment be characterized by pristine air or smoke-filled rooms? Here one notes that what counts as either a neutral environmental condition or a biological threat depends to a great extent on the characteristics of the organism. Generally, it is the characteristics of an organism that determine whether environmental conditions will be infectious, carcinogenic, etc. Again, it is very likely that there are significant bioengineering restraints involved in attempting to engineer a human organism able to be maximally resistant to all likely environmental challenges. One must make trade-offs. But trade-offs presuppose a way of comparing different morbidity and different mortality risks. To be authoritative, such comparisons presuppose a canonical value hierarchy, narrative, guiding principle, account of rightmaking conditions, etc., so that one can know how to appraise the value or rectitude of different trade-offs.

The matter is somewhat like comparing the worth of different currencies. Until one knows what the exchange rate is for Australian dollars, Canadian dollars, Hong Kong dollars, New Zealand dollars, and Singapore dollars, one cannot determine what one's reimbursement will be in American dollars for expenses incurred in these five currencies. So, too, one cannot calculate the consequences of different human biological designs unless one can compare the costs and benefits of different outcomes. In particular, one will need to know how to compare the risks of different morbidity and mortality losses (e.g., death early versus late in life). One must in addition compare present, near-term, and long-term costs and benefits. One must know the correct discount rate. One needs canonical guidance if one is authoritatively to discover rather than simply contractually or otherwise to authorize a particular policy.

The engineering design problem is thus compounded by the problem of comparing values. This challenge can be highlighted by continuing with the fantasy of redesigning the human organism and considering choices between increasing or decreasing the reproductive effectiveness of human intercourse so that individuals will generally have one or two children, or increasing the inclination to engage in non-reproductive "safe sexual" behavior rather than in reproductive sexual acts. Such engineering choices will in part reflect concerns regarding population growth and the transmission of venereal diseases, as well as the moral significance of "unnatural" acts. The recognition of such concerns cannot effectively lead to decisions unless one knows how to weight and compare the different costs and benefits at stake. So, too, if one considers the choice between making humans more inclined to live a healthy outdoor life to which they will be adapted versus making humans more inclined to live a healthy life of wandering dusty library shelves from which they will gain great satisfaction and to which they could in addition be well adapted, one requires a ranking of values to know which is preferable. So, also, if one considers the choice between adapting humans both physically and psychologically to outdoor activities versus a life closely bound to computers fully engaged in various virtual realities, one confronts different visions of the human project and of human flourishing.

Each of such choices and the myriad of others like them combines engineering design challenges with the problem of comparing the values of different notions of human flourishing. By invoking the fantasy of genetically reengineering human nature, one discloses substantively different senses of the normatively human that could guide different choices. The differences can be accented even further by tying them to the character of human finitude. In redesigning the human genome, there will be the possibility of not simply increasing life expectancy, but human lifespan. Such choices confront a human hunger for immortality as well as the finitude of human resources. To determine what trade-offs would be appropriate, and on what basis, between having individuals live much longer and losing the societal vigor that comes from more new individuals being born and entering into society, one must again have normative guidance. The longer more people live, the less reproduction there can be within any one period. Here one confronts not simply starkly different possible rankings of the values of outcomes, but choices among fundamentally different understandings of human finitude, the ultimate meaning of life, the character of transcendent reality, and the demands of God.

Though the example of germline genetic engineering is employed as a heuristic to underscore the divergent understandings of human flourishing that can guide different human genetic engineering projects, the diversity of values and the lack of a clear line between treating disease and enhancing function marks medicine generally. Circumstances are identified as states of disease or illness because they involve the absence of function, human grace, or human form considered to be appropriate, or the presence of inappropriate levels of pain and vexation, or the threat of a premature death. In ordinary circumstances without invoking the fantasy of genetic germline reengineering of the human organism, one confronts the problem of determining what dysfunctions, etc., should be the object of medical intervention. One is confronted with giving content to such cardinal terms as "appropriate", "inappropriate" and "premature". For example, if one is told of a patient who becomes seriously out of breath after running only a third of a block, one might consider that individual in need of significant medical attention. Yet if that individual is a 98-year-old man, one might have the sense that such circumstances should be accepted. After all, most men are dead at the age of 98. The problem of determining when species-typical decrements in function should be regarded as diseases exists throughout medicine. It includes considerations of when and if menopause, benign prostatic hyperplasia, or the loss of cardiorespiratory reserve associated with aging should be considered diseases. How much one treats and when one treats is guided by current expectations of what is feasible and appropriate. The terms "feasible" and "appropriate" cloak complex value comparisons and moral judgments.[4]

This is not to argue that one cannot generally agree as to what will count as diseases or illnesses across cultures and societies. Much of what is identified as a disease or illness is so termed, not because one agrees about the goals of human flourishing, but because such conditions are impediments to most human projects, most human undertakings, most notions of human flourishing. Even if there are substantial disagreements regarding human well-being, they usually do not matter, for such conditions (e.g., congestive heart failure, cancer of the pancreas, chronic obstructive pulmonary disease) constitute impediments for realizing nearly all understandings of the good human life, no matter how diverse. The point is that from the wide class of biological and psychological variations, some are selected as states of illness or disease because they are understood to be states associated with human suffering. However, human suffering is a highly value-infected, context-influenced phenomenon. For a state of affairs to count as a state of suffering, and in particular to be sufficient to merit medical attention, requires a judgment regarding appropriate levels of function, human grace, absence from pain, and avoidance of premature death. All of these states of affairs, however, can only be specified in terms of heavily value-laden understandings of human well-being.

This value- and context-dependent character of disease concepts was already well recognized in 1935 by Ludwik Fleck in his exploration of the history of syphilis and the ways in which its value-laden character has been constituted within particular thought communities with particular thought-styles (Fleck, 1935). Such thought-styles bring with them robust moral and non-moral value expectations, which are illustrated in the different ways in which both physicians and patients respond to the diagnosis of congestive heart failure on the one hand, and AIDS on the other, although both may have comparable mortality expectations. Particular notions of well-being, of proper human status, and of what constitutes unacceptable suffering, frame the full force of any diagnosis. Their particularity surfaces under circumstances that force the disclosure of the values involved. Otherwise, the circumstance that such states of affairs are serious impediments to the realization of most personal goals, so that they can be recognized as diseases without reference to particular values or notions of human flourishing, cloaks the different values and different notions of human well-being at stake.

The cardinal difficulty remains of identifying a canonical understanding of human well-being. Again, such cannot be disclosed by turning outward to nature, for in order to draw normative conclusions, one must already bring to nature a normative framework. This must be in hand prior to one's reading off moral implications from nature. Otherwise, one simply notes that there may be a range of socio-biologically-driven human inclinations, some to be peaceable, others to kill one's neighbor, others to get one's neighbor's wife pregnant (or alternatively, to become pregnant by one's neighbor's husband without his wife noticing, etc.).[5] To determine which are to be endorsed and which to be decried requires already having a view of what nature ought to be like. Again, the point is that in a post-Darwinian world one encounters wide variations of human biological and psychological traits. Some convey better or worse adaptation in particular environments. If the socio-biologists are correct, many are tied to strategies for maximizing inclusive fitness, which in many environments will be achieved by killing one's reproductive competitors, as well as engaging in complex strategies of seduction. As a consequence, what is "natural" in the sense of being part of the facts of the matter regarding how humans as organisms deport themselves, will not give moral guidance until one has prior moral guidance so as to sort out information from noise, endorsable activity from condemnable activity. One must bring to nature an interpretive framework so as to know what moral consequences one should draw from an examination of human nature.

The same difficulty attends the turn inward to human intuitions and sensibilities. One must already have in hand a particular thin theory of the good, moral sense, or moral rectitude to know which intuitions and sensibilities to endorse and in what order. Absent such a point of departure, one cannot determine from human or personal introspection what should be the canonical content of human flourishing. Nor can one ask what a disinterested observer would choose when confronted with divergent moral options, for a disinterested observer will choose nothing. Hypothetical decision makers, including hypothetical contractors, must already be fitted with a particular set of moral sentiments, a moral sense, or a thin theory of the good before they are able to make any particular choice. In short, one must already have a value framework or moral account in order to choose a value framework or moral account. One must already have what one was seeking in order to know what is normatively relevant.

As previously noted, this state of affairs will not be remedied by appealing to the consequences of different choices. One must first know how to compare consequences in order for a calculation of consequences to be possible. One must be able to compare impassioned versus rational preferences. One must be able to determine which is more valuable, the life of dispassion or the life guided by pleasurable engagement in strong passions. One must be able to compare the realization of future versus present preferences. One must know God's discount rate for time. Again, turning inward to an examination of the human person will not provide content-full moral guidance. The human person as a rational individual able to choose freely need not assign an overriding value to either rationality, life, or freedom. Without an appeal to a particular value-theoretical framework, all that one can establish of general secular moral significance is that persons are the source of bare permission as authorization. To gain content from an examination of the human person, one must already bring to that examination a normative understanding of the human person. One must also bring a normative understanding of the significance of rationality and freedom. But of course, this is what is at stake. As always in general secular reflections, in order to secure a canonical, normative content, one must either beg the question by granting initial crucial premises or engage in an infinite regress.

An examination of the human person provides no more moral content than a turn outward towards nature or inward towards the various intuitions and moral sentiments that individuals might possess. In every case, one is confronted by the need to know which rules of moral evidence and inference should be employed, and in what order to give weight to certain values, rightmaking conditions, and concerns regarding others so as to be able normatively to sort out information from noise. To discover a guiding, content-full moral vision, one must know how to choose among the different values and normative understandings one can bring to the moral appreciation of freedom, rationality, life, etc. As a consequence, one is again left with persons as the bare source of authorization, as the bare source of permission, as the origin of the authority of contracts and limited agreements. None of this will resolve the content-full controversies presented by health care.

III. THE NAKED PERSON AND THE DIVERSITY OF MORAL
COMMUNITIES

Recent debates in bioethics underscore the diversity of the moral visions at stake in health care delivery. Various different medical notions of well-being are inevitably placed in the context of substantial moral concerns. One finds divergent non-moral axiologies embedded in metaphysical understandings that provide substantial moral accounts of human well-being. Whether one regards a physician taking the life of a patient at the patient's request as euthanasia or as requested murder, whether one regards the age-old role of the physician as a terminator of pregnancy as one of providing treatment or as killing, whether one regards certain sexual inclinations as dispositions to alternative sexual lifestyles or to perversions, is determined by fundamental, framing understandings of the human project, morality, and the values that should guide human choice.

Well-being and human flourishing in health care are invested with deep moral significance. The rich moral differences that frame controversies in health care are embedded in webs of non-moral values and divergent construals of the deep structure of reality. These different webs of moral and non-moral values, as well as divergent metaphysical understandings, are for their part in great measure sustained by moral communities with divergent moral traditions and practices directed to sustaining different notions of human well-being. Such moral communities, as well as their traditions and practices, possess various degrees of robustness. Many, as Alasdair MacIntyre has observed (MacIntyre, 1981 and 1988 and 1990), are fractured and in decay.

This plurality of axiological and metaphysical understandings, the diversity of communities, and the shattered character of many moral practices cannot be bridged by a secular discovery of a canonical, content-full moral vision, which all should share. As has been shown, one must already concur regarding content-full moral premises and with respect to which rules of moral evidence and inference to employ in resolving secular moral controversies through sound rational argument. When one meets as moral and metaphysical strangers, the best one can do is to agree to tolerate each other's differences, collaborate peaceably where possible, and allow those who still maintain moral communities peaceably to pursue their own understandings of human well-being and flourishing. Which is to say, if one cannot discover a canonical, content-full philosophical understanding of human well-being that can be justified in general secular philosophical terms as obliging all, and if all do not convert to one true belief, then one can still provide authority through consent while tolerating diverse peaceable understandings of human flourishing, health, and well-being. The moral logic of a circumstance in which individuals have strong concerns regarding the nature of human well-being and flourishing, yet at the same time disagree regarding that nature, is that, if they are mutually tolerant, they will sustain different communities with alternative approaches to the realization of human well-being.

Disparate understandings of spiritual well-being have been pursued through parallel religious welfare systems, of which the contemporary German policy is probably one of the most instructive. Individuals committed to quite different understandings of church, sacraments, salvation, and grace employ a religiously neutral state as the tax collector to support two divergent views of spiritual welfare.[6] Rather than requiring a society to be united in a common religious vision, German society has instead come to understand itself as an open space within which two different religious communities can peaceably pursue their own understandings of the ultimate good. Health care would appear to be an area in which the toleration of differences in understandings of human well-being and flourishing would be easier than the toleration of religious differences. Though health care touches on many areas of ultimate concern, it does not have as many immediate ties to transcendent interests as does religion. At the least, it is likely that a similar agreement to tolerate those whom one holds to be fundamentally misguided could be established in areas of health care.

As one looks to a future in which it is unlikely that all will convert to a single understanding of the moral and non-moral values at stake in medicine, and in which competing understandings of well-being will continue to have a purchase on the health care interests of individuals, such a model can be used to guide persons in collaborating with others to sustain the diverse notions of well-being to which they are committed. Outside of a particular community with a particular moral vision, it will not be possible fully to realize the goods of health care. Health care goods are communal. Moreover, because of their ties to significant transcendent concerns regarding birth, copulation, and death, health care's concerns with well-being involve undertakings that both divide and unite (e.g., divergent views of the moral status of abortion and euthanasia, as well as different understandings of "unnatural acts" and human deviance). Those with substantial moral concerns will seek to maintain their moral integrity and to realize their content-full visions of human well-being. As most societies now allow differences in health care amenities, as well as differential access to certain expensive health care interventions, the future may very well witness the emergence of moral differences in health care provision.

The possibility of the emergence of parallel health care systems framed around content-full views of health and well-being will in any event in secular moral terms always be an appropriate possibility. This possibility has significant attraction, for the non-moral values that shape concepts of health and disease are embedded in complex moral and metaphysical concerns regarding the meaning of reproduction, birth, suffering, disability, and death. These concerns are elements of ways of life. They shape the character of moral communities. They are diverse. Given the broken character of the human condition, they are potentially legion. Medicine, because it addresses this whole range of concerns, brings these diverse understandings to bear in recognizing disease and health. The possibility of articulating divergent understandings and different communal approaches to the provision of health care underscores the diverse moral and non-moral visions at stake.

Outside of a particular context, a particular community, a particular history, a particular set of relationships, persons are naked. They have little meaning they themselves can establish as binding. It is within communities and their webs of commitments that persons come to participate in an intersubjective account of wellness or health. It is only within a communal context and its relationship that persons become clothed in taken-for-granted expectations and can see the content-full issues at stake in concerns regarding health and well-being.[7]

IV. PERSONS AND THE SPARSENESS OF SECULAR MORALITY

In all of this, one finds persons. Persons are central. They critically regard their own human nature and their relationships with other humans. Human nature is always a possible object to be judged and assessed by persons. Human nature is now also becoming an object to be refashioned in accord with the wishes of persons. In this respect, the Roman Catholic church is correct in understanding the revolutionary character of artificial contraception. Artificial contraception makes human nature plastic to the will of persons, at least in one area. Persons generally stand over against their nature, bringing to it diverse understandings of well-being. Which particular notions of well-being are appropriate becomes a matter of choice. This is the case with a wide range of choices regarding the nature of health, and likely will be ever more so, especially with the probable advent of germline genetic engineering.

Persons join with other persons in the pursuit of particular understandings of well-being. In the absence of content-full, canonical moral guidance, persons can still collaborate with the authority of permission. Indeed, considered outside of relations within particular moral communities, the ties that unite persons will be those of consent and agreement: consent to affirm particular understandings of well-being; agreement to collaborate within particular communities directed to particular understandings of well-being. From this perspective, particular content-full notions of disease, health, and well-being are the object of agreement and the concern of particular communities. Moreover, arrangements that use persons without their permission will not be justified in general secular terms.

Persons jointly fashion the content-full realities of disease, health, and well-being as social constructs. This is as one would expect. Disease language in medicine is often the conscious creation of international committees fashioning particular nomenclatures, gradings of disease, and stagings of pathology. In addition, shifting social concerns identify certain particular strands of causality as of interest in explaining the phenomena of pathology, not because they are the only strands, but because they are, from a particular perspective, the more useful or more socially acceptable. One might think, for example, of the phenomenon of HIV infection, which can as easily be regarded as a genetic disease (e.g., certain species of monkeys appear immune, and humans perhaps could be genetically engineered to have such immunity), as a social disease (e.g., dependent primarily on particular behavioral patterns), or as an infectious disease. In addition, as already observed, particular values identify "appropriate" levels of function, grace, freedom from pain, etc. Such usually non-moral values are brought to phenomena in order to draw lines between what will be considered pathological and what will be considered normal. Finally, classifications of diseases are tied to a rich web of obligations and rights, including disability payments, excuses from particular obligations, and rights to particular therapeutic interventions. To determine a state as one of illness, disease, health, or well-being is also performative. It places individuals in socially fashioned, value-infected roles. Any particular understanding of health, well-being, and disease depends on a complex nexus of both explicit and implicit social agreements.

The content-full reality of well-being and health is not from all perspectives and in all circumstances merely a social construct. Regarded from within a robust moral community, persons may experience their communality in a unity of moral vision and a concord in action, not seen as grounded in consent, but in the recognition of the claims of certain goods and truths, in particular, a common understanding of human well-being and human flourishing. Those who participate in a concert of right behavior and community, in accord with the depth of reality, may truly know well-being, but not in a fashion demonstrable to moral strangers. Well-being will be personal.[8] This is not unexpected if secular philosophy provides the logic for the collaboration of moral and metaphysical strangers in commonly constructed intersubjective realities, while theology is the experience of the deep personal root of all reality.[9]

Rice University
Houston, Texas

NOTES

[1] Chester Burns is one of the first to have recognized the plurality of concepts of health. See Burns, 1975.

[2] The literature concerning the value-infected and culture-dependent character of concepts of health, disease, and illness is rich and complex. There have been numerous attempts to defend the value-neutral account of concepts of health and disease (Boorse, 1975; 1977; Clouser et al., 1981; Kass, 1985). Others have defended the view that there are culture-neutral values integral to notions of disease, health, and illness (Von Wright, 1963). Yet others have advanced accounts that focus on action-theoretic accounts of the role played by vital goals in defining health (Nordenfelt, 1995). There has as well been a sustained defense of the value-infected and culturally framed character of concepts of disease, illness, and health (King, 1954; Margolis, 1976; and Goosens, 1980; see also the discussions in Nordenfelt and Lindahl, 1984). I have at length advanced arguments to show the unavoidably value-infected and cultural-relative character of concepts of disease, illness, and health (Engelhardt, 1996, chapter 5).

[3] As Henrik Wulff has argued, concepts of health, disease, and illness function as warrants for medical intervention (Wulff, 1981).

[4] These arguments are developed in detail elsewhere (Engelhardt, 1996).

[5] Sociobiological data and analyses suggest human "moral" inclinations are polymorphic and directed to maximizing inclusive fitness. As such, they involve numerous behaviors that are far from moral (Alexander, 1987; Symons, 1979; and Vogel, 1989).

[6] The author recognizes that the established Christian churches of Germany are in a state of profound crisis, given the desire on the part of many of their hierarchy and members for further novel refashioning of their religious commitments and beliefs.

[7] This observation is akin to Hegel's recognition of the difference between *Sittlichkeit* and *Moralität*. (See Engelhardt, 1994).

[8] If true well-being involves a relationship with God, and it is "impossible for human reasoning to disclose the truths of Christianity" (Vlachos, 1994, p. 23), then it will not be possible to discover a canonical content-full account of human well-being through a philosophical study of natural revelation, natural theology, or the human person. Theology is not study about God, but a relationship of prayer and communion with God. As Evagrios the Solitary (A.D. 345-399) stressed, "If you are a theologian, you will pray truly. And if you pray truly, you are a theologian" (Evagrios, 1979, p. 61). This is the case even with natural revelation, a point sorely missed in non-Orthodox Western theology. As Dumitru Staniloae observes, there is a recurring Western, non-Orthodox Christian mistake of holding that "in natural revelation man is the only active agent" (Staniloae, 1994, p. 17). From natural theology and natural philosophical reflections outside of a relationship to God of true worship, one may at best come to the paganism of the Greeks (p. 17).

[9] "God is not the Absolute Thou, but a living Person Who is in organic communion with man" (Vlachos 1994, p. 25).

REFERENCES

Alexander, R. D.: 1987, *The Biology of Moral Systems*, de Gruyter, New York.

Boorse, C.: 1975, 'On the distinction between disease and illness,' *Philosophy and Public Affairs* 5 (Fall), 49-68.

—— 1977, 'Health as a theoretical concept,' *Philosophy of Science* 44 (December), 542-573.

Burns, C. R.: 1975, 'Diseases versus healths: Some legacies in the philosophies of modern medical science,' in H. T. Engelhardt, Jr., and S. F. Spicker (eds.) *Evaluation and Explanation in the Biomedical Sciences*, D. Reidel, Dordrecht, pp. 29-47.

Clouser, K. D., Culver, C. M. and Gert, B.: 1981, 'Malady: A new treatment of disease,' *Hastings Center* Report 11 (June), 29-37.

Engelhardt, H. T., Jr.: 1996, *The Foundations of Bioethics* 2d ed, Oxford University Press, New York.

—— 1994, 'Sittlichkeit and post-modernity: An Hegelian reconsideration of the state,' in H. T. Engelhardt, Jr. and T. Pinkard (eds.), *Hegel Reconsidered*, Kluwer, Dordrecht, pp. 211-224.

Evagrios: 1979, 'On prayer' in *The Philokalia*, St. Nikodimos of the Holy Mountain and St. Makarios of Corinth (eds.), G. E. H. Palmer, P. Sherrard, and K. Ware (trans.): Faber and Faber, London.

Fleck, L.: 1935, *Genesis and Development of a Scientific Fact*, T. J. Trenn (trans.), University of Chicago Press, Chicago.

Goosens, W. K.: 1980, 'Values, health, and medicine,' *Philosophy of Science* 47 (March), 100-115.

Kass, L.: 1985, *Toward a More Natural Science: Biology and Human Affairs*, Free Press, New York.

King, L. S.: 1954, 'What is disease?' *Philosophy of Science* 21 (July), 193-203.

MacIntyre, A.: 1981, *After Virtue*, University of Notre Dame Press, Notre Dame, Indiana.

—— 1988, *Whose Justice? Which Rationality?* University of Notre Dame Press, Notre Dame, Indiana.

—— 1990, *Three Rival Versions of Moral Enquiry*, University of Notre Dame Press, Notre Dame, Indiana.

Margolis, J.: 1976, 'The concept of disease,' *Journal of Medicine and Philosophy* (September), 238-255.

Nordenfelt, L. and Lindahl, B.I.B. (eds.): 1984, *Health, Disease, and Causal Explanations in Medicine*, D. Reidel, Dordrecht.

Nordenfelt, L.: 1995, *On the Nature of Health*, Kluwer, Dordrecht.

Staniloae, D.: 1994, *The Experience of God*, I. Ionita and R.Barringer (trans.), Holy Cross Orthodox Press, Brookline, Mass.

Symons, D.: 1979, *The Evolution of Human Sexuality*, Oxford University Press, New York.

Vlachos, H.: 1994, *Orthodox Psychotherapy (The Science of the Fathers)*, E. Williams (trans.), Birth of the Theotokos Monastery, Levadia, Greece.

Vogel, C.: 1989, *Vom Töten zum Mord*, Carl Hanser, Munich.

Wright, G. H.v.: 1963, *The Varieties of Goodness*, Humanities Press, New York.

Wulff, H.: 1981, *Rational Diagnosis and Treatment*, 2d ed, Blackwell Scientific, London.

PATRICIA DONOHUE-WHITE
AND KATERYNA FEDORYKA CUDDEBACK

THE GOOD OF HEALTH:
AN ARGUMENT FOR AN OBJECTIVIST
UNDERSTANDING

I. INTRODUCTION

In contemporary analyses of the concept of health, it is generally undisputed that health in its primary meaning is a value concept[1]: health is a good and as such it motivates and regulates types of action. The good of health is further specified as a kind of well-being or as a necessary condition for well-being. The predominant understanding of good in general and well-being in particular is a subjectivist one (Nordenfelt, 1987, pp. 81-82), basing well-being variously in individual feelings of happiness or the satisfaction of desires (desire-satisfaction theory[2]).

Besides being the most commonly accepted framework, this subjectivist understanding of the good of health seems to carry the benefit of accounting for two important intuitions about well-being. These intuitions have motivated contemporary attempts to restructure medicine in a way that has persons, and not biological goals or mechanical structures, at its center. The first is that only a subjectivist account seems consonant with the actual variations among individuals concerning what constitutes their well-being (Nordenfelt, 1993; p. 279). The second intuition is the ever-growing conviction that health can only be adequately defined if the individual subject of health and her individual well-being are taken into account. This intuition is at the root of the dissatisfaction with and critique of previous models of health. These models understood health in terms of functioning and well-working (rather than well-being) and thus left out the specifically individual and personal dimensions of health. It seems that only a subjectivist framework does justice to the individual as individual. That is, only a subjectivist framework gives primacy to personal well-being, goals and desires, rather than to general biological species goals, and mechanical structures and functions.

With the individual person as decisive for what constitutes the end of medicine, it seems that a way has been found to overcome the mechanistic and impersonalist approaches to viewing health and the

P. Taboada, K. Fedoryka Cuddeback and P. Donohue-White (eds.), Person, Society and Value: Towards a Personalist Concept of Health, 165–185.
© 2002 *Kluwer Academic Publishers. Printed in Great Britain.*

practice of medicine. Identifying the foundation of why something is pursued or avoided through reference to specific individuals and their judgments or psychological states provides a basis for a 'personalist'[3] practice of medicine. This makes personal well-being normative for medical practice. Such a conception is personalist because it does not subordinate the individual person to species goals, to broader societal goals, or to any goals which she has not chosen herself.

We intend in this paper to take issue with this subjectivist paradigm for defining health. We agree that health is a kind of good that is essentially related to the good of well-being and we share the concern for establishing a more personalist approach to health care. However, we do not think that this concern requires a subjectivist notion of value and well-being. On the contrary, we think that an objectivistic framework does greater justice to our experience of value and well-being and also provides the foundations which the personalist motive requires. Furthermore, the desire-satisfaction model of good and well-being is plagued with difficulties that together constitute a serious argument against its being a coherent model for understanding the good at all. The inherent limitations of the desire-satisfaction model are recognized by most of its defenders. While there have been attempts to overcome these in various ways, it is our view that these problems have not and in fact cannot be overcome.

In what follows, we begin by presenting the desire-satisfaction model of good and well-being, examining the attempts to overcome its problematic character, and bringing out what seems to us to lie at the root of the ultimate impossibility of solving the problem. In the second part of the paper, we present schematically an objectivist approach that is not only theoretically coherent, but which also does greater justice to our experience of the good of health. This approach is capable of grounding a rich personalist understanding of both health and medicine.

In discussing both the subjectivist and the objectivist alternatives as approaches to defining health, we begin not by directly considering specific competing definitions of health, but rather by analyzing competing notions of the good and well-being. Our disagreement with the definitions of health currently described as "normative" is not on the point of defining health in terms of the good. On this point, we are in fact in agreement with, for instance, Nordenfelt, Whitbeck and Pörn, and in disagreement with Boorse. Our disagreement focuses rather on the particular understanding of the good underlying the so-called 'normativist' (i.e., value-based) definitions of health. These considerations form the basis for our proposed definition of health.

II. DESIRE-SATISFACTION THEORY AND THE GOOD OF HEALTH

Desire-satisfaction theory explains the good not only in relation to an individual person, but, further, in strictly individual, subject-relative terms. Such terms may vary, being either happiness, desire-satisfaction, goal-achievement, etc. When we use the term "subject-relative," we mean that the good is understood in reference to an actual psychological state of a subject, which by definition must be individual. When we say that such an account understands the good 'in terms of such states', we intend to leave open the possibility for a subjectivist theory to claim that the good either consists wholly in individual psychological states (such as hedonism would claim),[4] or that the good depends importantly on such states, in that it is fundamentally a function of such states (as is the case in desire-satisfaction theories of the good) (cf. Brink, 1989, pp. 220-1; Kraut, 1979), though in what follows we treat only of the second type.

In this second kind of subject-relative conception, the good is not simply identified with subjective states such as happiness or desire-satisfaction. Rather, the good is defined as something which contributes to or is a condition of such a state: a thing is good if it satisfies a desire, or contributes to this satisfaction.[5] While not only personal, psychological states are good, they are decisive for whatever else is good. Hence, the good importantly depends on, and in some fundamental sense is a function of, individual psychological states.

A common and obvious objection to defining the good in reference to individual psychological states is grounded in the recognition that such a theory would in many cases force us to accept as valuable or good something which we intuitively would claim is not at all significant, or is even a disvalue. Such a theory would force us to admit, for example, that the satisfaction of a Nazi's desire to persecute non-Aryans is valuable, or to move to the absurd, that achieving the smallest handwriting as a life goal is valuable, if so desired.[6]

In response to this, many subjectivist theories introduce various 'objectifying' criteria, which limit the kinds of things which could be considered good. This is most commonly done by introducing the criteria of rationality coupled with the notion of 'ideal desires', i.e., the desires an individual would have were she fully informed and rational in her choices (cf., e.g., Railton, 1986, and Brandt, 1979). 'Ideal desires' are identified through reference to what would constitute being fully or adequately 'informed' and fully or adequately 'rational' with regard to the choices we make and the things we desire. Such criteria are called by the desire-satisfaction theorist "informational" and "motivational" constraints (these will be further discussed later). Typical to both is the assumption that knowledge affects desires or attitudes, and that a change in knowledge will effect a change in motivation.

With such an addition, subjectivist theories do not understand value or the good as merely a function of any subjective state. They understand the good, rather, as a function of a subjective state which is proper to a subject who is fully informed about a situation, and who acts or reacts in any particular case in a way consistent with this information, and with her other desires, plans or attitudes, etc. Recognizing the incompatibility between one desire and others, or between a desired object and the given conditions for attaining it, recognizing the factual inability of a given object to secure a desired subjective state, etc., provide sufficient grounds for evaluating our own or another's state. It also provides the basis for establishing criteria on the basis of which we can judge a given state as rational or not. Thus, the fact that someone claims that the satisfaction of a certain desire or the achievement of some end will make this person happy, does not force the subjectivist to conclude necessarily that this desire or this end is a good.

A subjectivist theory that incorporates such factors is, obviously, significantly different from one that does not. It seems that it is not 'forced' into the same counterintuitive conclusions that a purely subjectivist theory must accept. Nevertheless, as a desire satisfaction theory, it still identifies subject-relative states as the fundamental determination of the good. Therefore, it is still fundamentally a subjectivist explanation of the good, despite its incorporation of criteria directed towards the exclusion of arbitrary or irrational factors.

Within the discussion of health, well-being is understood as a kind of good – the good of 'being well'. If health is understood as the 'being well' of an individual, then it is consequently a value-based concept, because the notion of the 'good' is part of the very concept of health. The good of well-being, when understood in terms of desire-satisfaction, identifies individual well-being as that state in which none, or no significant amount, of the desires of an individual remain unsatisfied.

Because it is relative to individual desires, well-being is also relative to individuals, is subject-specific and therefore 'personalistically' conceptualized. It is this subject-specificity which gives it its primary recommendation for a 'personalist' reform of medicine.

Both Caroline Whitbeck (1981) and Ingmar Pörn (1984) have introduced theories of health which understand the good of health in these terms. What health is and the value it has are determined in terms of whatever contributes to the fulfillment of self-set goals. A goal, furthermore, is something a person wants to become the case. The good of health, therefore, lies in the way in which it is able to contribute to the satisfaction of individual wants.

While standing in fundamental agreement with this approach, Lennart Nordenfelt has pointed out the need to improve upon these theories, because they lead to the same counterintuitive consequences that we found inseparable from an actual desire-satisfaction theory of the good. Nordenfelt identifies health as the set of causal preconditions for that well-being which is itself the set of minimal conditions for happiness, where 'happiness' refers to well-being in the broader sense (1993b, p. 35). This narrow sense of well-being which is the good of health can also be specified in terms of 'vital goals' which Nordenfelt defines (1993b, p. 96) as "a state which is necessary for [a person's] minimal happiness." And to make the analysis of all the terms complete, Nordenfelt explains happiness as an equilibrium between an individual's wants or desires and the satisfaction of these wants (1993b, p. 45).

In other words, Nordenfelt's conception of health identifies health as that set of abilities identified through their instrumentality for satisfying those desires which constitute an individual's minimal happiness, or minimal well-being. The introduction of the word "minimal" serves to delimit the abilities of health from other capacities. It does not alter the fundamental meaning of well-being, which is still defined in reference to happiness. Whether a person is in health, therefore, depends upon what is necessary for the happiness of this person. And this, further, depends upon what constitutes the set of her wants.

We thus arrive at a conception of health that is subject-relative in precisely the way that we analyzed above. As with the general subject-relative notion of value presented above, health is fundamentally dependent upon individual wishes and desires, and varies with individual wishes and desires. Where Nordenfelt differs from Whitbeck and Pörn is that he includes the possibility of 'constraints' on individual goals, in order to avoid the difficulties of an actual desire-satisfaction foundation for the minimal conditions of well-being (Nordenfelt, 1987, pp. 76-96). He thus introduces factors aimed at resolving the problematic character of, for instance, goals set under compulsion, counterproductive goals, or the unrealistic goals of an over- or under-achiever. Because Nordenfelt's notion of well-being can be specified in reference to ideal rather than actual goals of individuals, it seems better able to account for the "objective" character of well-being, i.e., the fact of objective evaluation which seems to be a necessary element of clinical evaluation.

III. THEORETICAL DIFFICULTIES OF DESIRE-SATISFACTION THEORY

As widely accepted and as well worked out as desire-satisfaction theories of the good may be, they have been criticized on the basis of problems which seem to us insoluble and ultimately destructive of the theory as a whole (Cf. Brink, 1989; and Rosati, 1995). In summary, the difficulties amount to the impossibility of avoiding the counterintuitive conclusions necessitated by an actual desire-satisfaction theory without introducing factors that in the end force us wholly to prescind from the subject and her actual desires in the identification of her well-being and her good.

In order to exclude such counterintuitive notions of the good or well-being (such as the Nazi's desire to kill non-Aryans or the desire for the smallest handwriting), the desire-satisfaction theorist must deny the fact that it is the desires and wants of the subject which primarily and fundamentally determine well-being and the good. But this is the fundamental premise of the desire-satisfaction theory. Thus, the desire-satisfaction theorist must either opt for theoretical consistency and thereby allow that the desires and goals of, e.g., the Nazi and the woman desiring the smallest handwriting, are equally valid and determining of the good, or introduce other determining and evaluative factors according to which we can exclude desires/goals of these kinds. But in introducing these other factors, the fundamental premise of the theory is undermined, i.e., that the individual psychological states of the subject primarily and fundamentally determine what constitutes her well-being and her good.

There is a further difficulty concerning the relation between knowledge and motivation: even if one could acquire the fuller knowledge required to determine ideal desires, it is not evident why this knowledge would alter the subject's actual desire or become normative for the actual subject's action. There obviously are cases in which desires persist even in the presence of 'better judgment' and knowledge. The alcoholic is a ready example of one whose desire for alcohol does not disappear simply because the consequences of its abuse are made clear to her. The problem here is not a lack of knowledge, but what could be called a lack of rationality or of the ability to be properly motivated by this knowledge.

One way of countering this difficulty is by incorporating an ideal motive along with the knowledge: the ideal according to which the actual motive is evaluative includes both the subject having full knowledge and being rightly motivated by such knowledge. In this case, however, it is no longer the desire which is being determined (by the necessary information), but the subject herself, who is to respond to this information. In other words, it is not just that the desire is placed in a hypothetical situation, e.g., of more perfect knowledge, but it is placed into a completely hypothetical subject, e.g., a subject who recognizes the relevant facts, understands their implications for her life, prefers long-term happiness to immediate satisfaction, and in consequence desires those things which correspond to this long-term happiness. This subject furthermore does not desire anything that is contrary to, or would inhibit this long term happiness (Brink, 1989). However, though we may imagine a subject in the hypothetical situation of being fully informed and ready to make the 'rational' choice based on this knowledge, there is no guarantee that this 'ideal, fully informed' subject is someone whose judgments the real subject will consider normative when determining her real desires. In other words, the introduction of these informational and motivational constraints results in a hypothetical subject who is in fact no longer the real subject whose desire is now to be determined. A hypothetical identity between the two subjects does not ensure that the real self will recognize her 'ideal' self as authoritative. Assuring this would require the introduction of still further constraints, removing the moment of evaluation still farther from a real subject and a real situation (cf. Rosati, 1995).

Contemporary definitions of well-being in terms of self-set goal-attainment, need, or desire-satisfaction fare no differently than this general account of the good (cf., e.g., Pörn, 1984; Whitbeck, 1981; Nordenfelt, 1987). Nordenfelt does, as we have seen, develop a set of criteria or constraints for evaluating individually set goals and excluding the irrational from this set. This assures us, he writes, of a way of dealing with 'unreasonable' goals and wants, and gives a certain character of 'objectivity' and 'objectivizability' to his notion of well-being. Nordenfelt admits, however, that in the final instance, no amount of conceptual analysis or theoretical consideration can decide what qualifies as a condition for happiness in a given instance. As he expresses it, "the question of what constitutes 'real' happiness, as well as a minimal degree of such, can only be answered by a primary evaluation" (1987, p. 78). In other words, the individual alone is the only true judge of her own happiness. What factually contributes to this happiness is open to second-person evaluation, but since the 'pole' in reference to which this evaluation must occur is first-person determined, what objectively constitutes a condition for this happiness is also individual judgment dependent. While rationality criteria may be introduced with certain evaluatory functions, this notion of well-being remains at root "*egocentric*", and the function primarily pedagogic or heuristic. Because of this, there is no real or normative way to dispute a counterintuitive, counterproductive, or unrealistic view of what will satisfy a person's wants and therefore be constitutive of their well-being.

IV. AN OBJECTIVIST INTERPRETATION OF THE GOOD OF HEALTH

The objectivist position which we intend to articulate here also understands the good of health in terms of the broader category of well-being. Unlike subjectivist accounts, however, we affirm both the objectivity of good (or value) as such, and an objective meaning of well-being that is derived from the nature of the being in question. In order to elaborate this position, we will begin with a number of conceptual and terminological clarifications. We restrict the notion of well-being in its primary sense to living beings, i.e., beings of whom it is meaningful to say that they thrive and flourish, that they are well or ill (cf. Seifert, in this volume and v. Wright, 1993, pp. 50-51). When some being is sufficiently actualized, the being is said to thrive, prosper or flourish. In other words, the being is said to *be well*. Though it is meaningful to speak of the well-being of all living beings (and to speak of the health of all living beings), in what follows we will narrow our focus to an analysis of the well-being of human persons.

In the context of an analysis of well-being, we mean to distinguish two phenomena designated by the term "good."[7] The first phenomenon is the comprehensive, fundamental good of some being. In what follows we refer to the actualization of this comprehensive good (*the* good of some being) with the term well-being. The second phenomenon consists in the various particular things (e.g., objects, activities, ends) that are good *through their beneficial relation* to the well-being of some being. We refer to such things as *beneficial goods*.[8] The category of the beneficial includes, to use von Wright's words, everything which "affects favorably" the good of some being, which "serves the good of the being," or which is "promotive of that peculiar good which we call the good of the being" (Wright, 1993, pp. 43, 45).[9]

When we speak of the good of health, we often speak of its beneficial character as though health were simply a beneficial good: health is good for some being because it serves or promotes well-being. We suggest, however, that the good of health is not simply a beneficial good as are, for example, exercising regularly and eating well. It is also and most properly a type of well-being which is an integral dimension of the more comprehensive well-being. It is thus both well-being itself (by being a part of comprehensive well-being) and beneficial in the sense that it serves or promotes the general, comprehensive good of the being.

The objectivist understanding of well-being we defend can be stated negatively in terms of the desire-satisfaction theory: actual well-being does not consist merely in the satisfaction of desires, in the ability to satisfy desires, or in an equilibrium between basic desires and their satisfaction. Two claims can in fact be distinguished here: (1) the well-being of a person is *not good through being desired*, actually or ideally, since value is not merely a function of desire; and (2) *what* the well-being of a person consists in is not determined by her desires, even those desires which are considered basic, fundamental, or 'goal-establishing' for her life. The first concerns the specifically value-dimension of well-being, what it means to say that the well-being of some being is *good*, in the general sense of saying that it has value or is a value. The second claim concerns the 'content' of well-being, or in what the well-being of some being consists. We intend to articulate an objectivism in both cases.

V. THE OBJECTIVITY OF VALUE

We begin with the objectivity of value. To say that well-being is not good through being desired is to say that its value-dimension is not merely a function of desires, actual or ideal, or more generally, that value itself is not merely a function of desire (where 'function' means constituted by or through). In saying that the value-dimension of something is not constituted by or through a person's desire for that thing, we mean first of all to establish the *causal independence* of the value-dimension of the thing from any *individual* and *actual* desire for that thing (desire here understood as an experienced psychological state or inclination). One can point to various typical experiences in which value and desire are at variance, and which illustrate this independence. A person can know that something is good or of value and yet not desire that thing or even be repelled by it. A person could know, for example, that studying philosophy is good, a specific work of art is good, or that it is good to tell the truth, and yet have no desire to study philosophy, contemplate the work of art, or tell the truth in this instance. In the opposite case, a person can know that something is not good and yet nonetheless desire it: a person could, e.g., desire to drink alcohol excessively or hurt another person. There are also other types of value experience in which it is meaningless to speak of desire playing any role at all. In marveling at the beauty of the night sky for example, there is no meaningful sense in which one could say that one has any kind of desire for the night sky (Scheler, 1973).

From the claim that value is not a function of desire, it does not follow that there are no meaningful relations between the value-dimension or good of something and the desires of personal subjects. Desires are typically related in a meaningful way to the value-dimension of things; we typically desire things in virtue of their value. What we mean to establish here is the causal independence of value in relation to desire and also the priority of value in relation to desire. In other words, we mean to establish the *direction* of the relation between desire and value: desire (when rational) is typically *motivated* by the value-dimension of the thing. This implies that the value is in some sense prior to the desire; it is not constituted through being desired (Scheler, 1973; Butchvarov, 1989).[10] More significantly, we generally judge whether a desire is rational, ordered, or moral in terms of its adequacy to the value of the thing desired. A deep and long-held desire to achieve the smallest handwriting is disordered because such a goal is trivial, where trivial is a value term which characterizes the axiological dimension of such a goal. The Nazi's desire to eliminate non-Aryans is evil and morally disordered in the strict sense. Thus, the priority of value in relation to desires also means that desires are judged in light of values and not the other way around.

Actions can also be motivated by the value of something independently of any desire for that thing. Examples of conflicts between desires and actions motivated by value illustrate this point. A person could find exercising daily painful and yet do it because it is beneficial to her health. The fact that she desires to be healthy and knows that exercise is instrumental to the good she desires does not alter the fact that she does not desire to exercise. Here there is an actual conflict between her lack of desire to exercise, or, more strongly, her aversion to it, and her choice nonetheless to engage in that activity. In the very different case of non-instrumental goods such as moral values, we are often motivated to act against our desires. Confronting and speaking out against injustice or openly refusing to cooperate in evil can be extremely painful and difficult and can result in the loss of other things one values, e.g., friendship, respect, reputation, professional success, security, even health and life in some cases. To say that in each of these cases one is motivated by some deeper desire, e.g., the desire to be just, is not to the point. Even if there is such a deeper desire, it is not the primary motivating factor in this instance. Rather, it is the value or disvalue itself which motivates primarily and often against one's deep desires for security, respect, friendship, and life.

To this point we have been stating our objectivist position negatively in terms of the desire-satisfaction theory: value is not merely a function of desire. We can now offer a positive formulation: to say that something is objectively good or has objective value is to say either that it is good *through itself* or that it is good through its objective relation to some other good. To say that it is good through itself is to say that it is not constituted as good through any act or psychological state of a subject. Here then we extend the meaning of objective to include a causal independence not only from the desires of some subject, but from any acts of a subject or any subjective psychological states. Objective goodness or value is thus an objective quality or dimension of a thing which that thing possesses independently of any acts or psychological states of a subject. The type of good which we refer to with the term 'beneficial good' however, is not good simply through itself. Rather, it is good *through* its beneficial relation to some other good. A beneficial good objectively serves or promotes some other good.[11]

We want to suggest that well-being (the state of being well) is both good through itself and beneficial. Well-being considered comprehensively and *qua* good, is good through itself. However, well-being is also a complex good; each of its dimensions or components is also a beneficial good since it serves or promotes the general, comprehensive good. One could say further that as the well-being of some subject, well-being is essentially relational; as the good *of* some subject, it is also good *for* that subject. In both cases, the value-dimension is objective in the sense that it is not a function of acts, inclinations or psychological states of a subject. Being healthy, we argue, is an integral dimension of well-being. As such, it shares in the good-through-itself character of well-being. As one component of well-being, however, being healthy can also be said to be a beneficial good since it promotes the general, comprehensive well-being of the subject.

VI. OBJECTIVE DIMENSIONS OF WELL-BEING

We turn now to the second claim of the objectivism we are defending: the content of well-being is not determined solely or simply by the subject's desires, even those desires which are considered basic, fundamental, or 'goal-establishing' for her life. In order to proceed, we suggest distinguishing two phenomena referred to by the term 'well-being': an experienced state of *feeling* well and a state of *being* well. The first is primarily a psychological category, the second an ontological category.[12] Both phenomena are objective in the sense that they are real, actual phenomena.[13] 'Feeling well' refers to a consciously experienced *feeling state* of a subject. Being consciously experienced is essential to such a phenomenon; it is meaningless to speak of a feeling state that is not consciously experienced. In contrast, 'being well' refers to an actual *state of being* which may or may not be experienced. A state of being is an ontological category and as such it is not constituted as or through conscious experience. This does not, however, mean that it cannot be consciously experienced or that it is not normally experienced. We would argue that there is naturally a coincidence between the two states – 'being well' is naturally experienced as 'feeling well' – but this coincidence does not undo their distinction from one another.

Both states can also be said to be subjective, though in different ways. 'Feeling well' is subjective in the primary sense, where 'subjective' means consciously experienced by a subject. 'Subjective' in this sense does not mean arbitrary, unreal or unintelligible. It is the broad category of 'the consciously experienced'. 'Feeling well' is a type of subjective experience which is a self-experience, i.e., an experience of oneself, as distinct from intentional experiences of other things. 'Being well' can also be said to be subjective in a secondary sense; it refers to a state *of* a personal subject though it is not constituted as or through conscious experience.

'Being well', as we have said, is normally experienced as 'feeling well'. However it is not reducible to a feeling state and, more importantly, it is foundational for 'feeling well' when 'feeling well' is objectively grounded. In other words, 'being well' is not a function of 'feeling well', and the latter is normally consequent upon the former. Instances of persons 'feeling well' and yet not 'being well' illustrate this distinction, e.g., a person with breast cancer who has as yet no experienced symptoms, or a person who represses and denies an experience of extreme violation or loss. In both cases, the person may say that she feels well while in fact she is suffering, in the first case from a life-threatening illness and in the second from a deep psychological wound. While a person could 'feel well' and yet not 'be well', we do not mean to affirm that one could just as easily 'be well' and yet not 'feel well'. If a person does not feel well, physically or psychologically, then something is 'wrong' even if there seems to be no bio-medical cause or recognizable syndrome.[14] In our usage, the term 'well-being' never refers merely to a feeling state, but rather and primarily to that state of being which underlies and grounds a state of 'feeling well' when the latter is objectively grounded.

According to desire-satisfaction theory, it is the desires of the subject which determine the content of well-being, since well-being consists in the satisfaction of desires or an equilibrium between desires and their satisfaction. Well-being is achieved or promoted when desires are satisfied or fulfilled, or when the satisfaction of them is furthered, whether this satisfaction is understood to consist in the actual realization of a goal (the object desired is gained) or in the experience of satisfaction, i.e., the experience of a specific feeling state.[15] Articulating our position in opposition to this, we can be more specific about what we mean by the objectivity of well-being. We mean to say that the content of well-being is not determined solely by the desires and choices of the subject, even those desires and choices which are fundamental and goal-establishing.

From the fact that it is typical to desire and choose those things which promote one's own well-being, it does not follow that what constitutes that well-being is determined by desire or choice. In many cases, desire, and more commonly choice, follows on a comprehension of what is promotive of well-being (even where that comprehension is mistaken). A person typically chooses a goal or end precisely because she thinks it will promote her well-being and, whereas deeply and long held desires are often indicators of what constitutes personal well-being, they, along with choices, can also be at variance with well-being.[16] A person could, for example, unknowingly choose (and desire) a career which does not suit her abilities and thus be continually frustrated, or she could engage in activities that she knows will undermine her health or her moral character. A person could even cease to desire her own well-being altogether as in the case of suicide motivated by self-hatred or deep depression.

It is not unusual to experience a conflict between a desire for *desire-satisfaction* understood as a feeling state and what one knows to be one's general well-being. A person could find e.g., eating and drinking excessively quite pleasurable and yet know that such behavior undermines her health and thus her general well-being. Often we can identify certain activities or goals as good for us in the sense that they promote our well-being and yet we take no pleasure in them or even find them deeply painful or distasteful. In such situations, we often need to turn to a consideration of our general well-being in order to be motivated to perform the activities or achieve the goals.

One thing that gives weight to the idea that desires determine well-being is the fundamental intuition that there ought to be a correspondence between our deeper, long-held desires and our well-being. Intuitively we grasp that 'something is wrong' when a person does not desire her own well-being (as in the case of a person suffering severe depression), when a person deeply desires something trivial or superficial (e.g., desiring the smallest handwriting), or when a person deeply desires something morally abhorrent (the case of the committed Nazi). If we examine this intuition closely, however, it is clear that well-being cannot simply be constituted by desires. In each of these cases we want to say that the desire is disordered, inordinate, or immoral and this implies that the desires of the subject are not determinate of her well-being.

If the individual desires and choices of the person do not determine the content of her well-being, what does? We would argue that it is the nature of the person that fundamentally determines well-being.[17] In saying that nature *fundamentally* determines well-being, we mean to indicate that the concrete, specific well-being of an individual person is not solely and exhaustively determined by nature. There are other factors such as essentially individual traits and abilities, individual choices and desires, and external factors such as social structures, culture and environment which either partially determine the form and/or content of the well-being of a particular person, or modify well-being in various ways. These factors, however, are always 'in the background' of nature and within the limits of nature. Our claim is that the primary, general determination of human well-being is human nature. Thus, we mean to affirm that there is a foundational and generic human well-being which follows from the general structure of human nature. We will consider some of the individual factors which partially determine and modify well-being below. At this point, we want to clarify the concept of nature.

With the term 'nature' we mean first of all to refer to a discernible, general design or structure that determines an individual as a member of a living species. To use Boorse's terminology, we mean to refer to a 'species design' and to the development and function proper to that design (Boorse, 1975). In contrast to Boorse, however, we would affirm that this design is a natural kind (not merely a statistical norm) discernible through observation and investigation (see Kass, 1985 and Greene, 1978). The point, in its most general and minimalist form, is simply that nature "has somehow got itself sorted into kinds, these kinds turn up in their individual instances, with fair but not total regularity, and the exceptions deviate somehow from the standard embodied in the 'best' case" (Greene, 1978, p. 130).[18]

In speaking of natural kinds and design, we mean to refer to a type of normativity – to what Kass calls a *natural norm*. A natural norm, to follow Kass, is "not a moral norm, not a 'value' as opposed to a 'fact', not an obligation" (Kass, 1985, pp. 173-74). To speak of a 'natural norm' is to refer, first of all, to an objective norm, i.e., a norm discovered in nature. As various thinkers have argued, this concept of normativity is quite commonly and necessarily employed in scientific processes of classification. In biology and zoology, e.g., the identification and classification of organisms presupposes some recognition of natural kinds and of normal development and functioning (e.g., Kass, 1985; Greene, 1978; Wallace, 1978). Proper development and functioning, in turn, presuppose some concept of inherent *telos* (e.g., generation, self-organization, self-maintenance, self-preservation) (Kass, 1985, p. 101). Natural norms are understood to incorporate ranges of variation, and are subject to modifications due to, e.g., environmental conditions. Thus, the objectivist position we defend does not hold that such natural norms are necessary or absolute, but rather, that they are sufficiently stable to allow for their specification within a certain range.[19]

It is against this background that we introduce and understand the notion of the "good of a being" – what is often called "well-being," or "welfare." In his *Varieties of Goodness* (1993, ch. 3), von Wright introduces a type of good which he calls the "welfare" good, or "the good of a being." As already mentioned, von Wright felt that only living beings can have "a good." It makes no sense to speak of the good of a stone, but it does make sense to speak of the good of a tree, and more clearly, of a person. Von Wright roots the possibility of "having a good" in life, in the dynamic kind of being which is life.

When speaking of the good of persons in particular, von Wright speaks not primarily in reference to the life of persons, but in reference to their wishes and desires. We do not disagree that there is a sense of the good of persons which is related to their desires or needs. We want, however, to elaborate a notion of the good of a person which is rooted in the part of that person which is not subject to or a product of her acts or desires. We will do this in terms of the way in which Stephen Toulmin analyzes the intrinsic "evaluative" character of organic function. According to Toulmin (1975), the *end* which establishes function as function is a "value" – something that is striven for by the function itself. With regard to health – and this is where we begin introducing our proposal for a definition of health – the end, or the good to be achieved, is the function which is the *actualizing* of the whole itself. This actualization, moreover, is the specific actualization of the being as a *natural* kind, as elaborated by Kass and Greene, and not as a free individual (who makes choices and has desires).

What does it mean to say that the flourishing or actualizing of a nature is a good? A good that can be promoted is a living good. It is a vital existence that unfolds its existence in time. It is an existence that can be well or not well, flourish or decay. These are the marks which Marjorie

Greene identifies as proper to natural kinds – or natures in the Aristotelian sense (1978). The criterion according to which we recognize whether the development of a being constitutes a flourishing or a decaying – and is hence a being-well or not – is its own nature.

What we have here is our disputed notion of "well-being" or "welfare" objectively understood in terms of the flourishing of a nature, which is a good of a being. What constitutes the well-being or actualization of the organic entity in this sense is determined in reference to its nature, in the Aristotelian sense. Thus, this notion involves taking Toulmin's notion of "function" and enlarging it to mean "natural function," or, function of a natural kind, in Grene's and Kass' sense of the term

Of course, when we consider a human being, we are not concerned only with the human being as a member of a biological species, and we do not treat simply of a general nature which organically unfolds or is actualized as we do with plants, and to a different degree with animals. We also consider the human being as a person and therefore introduce other dimensions which are essential to personal life and well-being: the realms of rationality, affectivity and freedom, the realms of meaning, values, morality, and spirituality, and the realms of social and interpersonal relations. To these various dimensions of personal life correspond dimensions of well-being, e.g., physiological, psychological, social and moral, and their various interrelations. Because the human person is so complex, well-being can be specified in terms of different contents depending on the dimension of the person which is emphasized and the approach to the person which is employed (we have in mind the vastly different approaches of, e.g., bio-medicine, psychology, philosophy, sociology, theology, etc.). This manifoldness, however, does not undermine the objective dimension of well-being, nor its generalizability. It simply means that comprehensive, personal well-being defies simple definition.

A further limitation on forming a comprehensive definition of the human person and, consequently, personal well-being arises from the individuality of personal subjectivity, which is not at all explicable through variations in species design. In particular, with human persons we introduce the notion of freedom and, thus, an inherent creativity and "unpredictability" (Mounier, 1952).[20] The person has the capacity to actualize or frustrate her own potentialities through action and choice, and to determine to a large degree her own goals and purposes in life. The person's particular abilities, her actions and choices, and her likes and dislikes will influence and modify the particular content of her well-being. In a similar fashion, her desires may determine which dimensions of her person she chooses to actualize and the way in which she actualizes them. As with the manifoldness of the person, personal individuality and freedom do not negate the significance of the generality of human nature and the fundamental, general well-being which is consequent upon it. Human nature forms the 'background' of individuality and freedom, and generic human well-being, while it is necessarily varied in its concrete, specific forms, remains foundational for individual well-being.

To summarize our general point: well-being can be specified in general terms as the actualization of the *good of the human person*. The basic character of this good is determined by the general nature of the human person modified through its instantiation in particular human subjects. Well-being in the full and comprehensive sense is further determined by the unique potentialities of the individual (capacities, abilities, characteristics), and, more importantly, by the actions and choices of the individual person. Nonetheless, generic well-being remains foundational for this fuller, comprehensive well-being. The human person thrives or flourishes when her well-being is actualized and an experience of fulfillment or satisfaction, i.e., the experience of 'feeling well', typically and even properly accompanies actualization. The phenomenon of a deeply felt fulfillment (happiness) is a natural consequence of the actualization of well-being and an integral dimension of well-being itself.

Health, understood in this framework, is a specific relation between the good of a being (welfare or well-being) and the nature of that being. Health is a good, and, more particularly, it is a dimension of the good of a being which is a direct function of the nature of that being, nature here being understood precisely in distinction to freedom. Through this, health is limited to that sphere of actualization that happens not in virtue of an individual's free acts or desires. Rather, it is a dimension of the individual in which a certain unfolding of being happens in virtue of that being's nature. It is good, it is good for a being, and can be desired by a being. It and its goodness cannot, however, be reduced to simply being desired.

VII. CONCLUSION

We believe that the objectivist position we have articulated is more adequate to a personalist approach than a subject-relative approach, which allows any desire, even those intuitively harmful for persons, to be constitutive of their good. It recognizes and affirms that the human being is a person and not merely a member of a biological species. It takes with full seriousness the realm of conscious experience: the person's experience of herself, her body, mind, affective life, personal relations, relations to the world around her, and the integrity (or loss of integrity) of these different dimensions. It recognizes the significance of personal freedom and responsibility and the centrality that values play in personal life and action. Since both the patient and the healthcare provider are persons, considerations of value must be at the center of healthcare practice.

Within an objectivist framework, however, the recognition of this fact does not imply theoretical relativism or lead to the acceptance of conflicts which cannot in principle be resolved. The task of reaching understanding and agreement on matters of value is, without doubt, often arduous and contentious and, in the case of serious moral matters, often unsuccessful. But an objectivist can be committed to engaging in the process precisely because she believes that understanding and agreement are possible and therefore eminently worth working and struggling for. Our hope is that this paper will contribute to this goal.

International Academy of Philosophy
Principality of Liechtenstein

NOTES

[1] Cf. Engelhardt, 1976, 1981; Nordenfelt, 1987; and Whitbeck, 1981. The notable exception to this is the well-known position of Boorse (cf. 1975, 1977).

[2] We refer to the general theory as 'desire satisfaction' though it is also referred to as, e.g., preference satisfaction, or need-satisfaction. The differences among these accounts are not relevant to our analysis since we are concerned with the general thesis that the good is defined fundamentally in terms of individual psychological states.

[3] Although the term 'personalist' has within certain philosophies a more precise and more extensive use and meaning, it seems to be used within the discussion of health to point to the concern of doing justice to the individual *qua* individual. It is in this sense that we use the word here.

[4] A theory that understands the good wholly in terms of subjective states would be for instance the psychological hedonism of J.S. Mill, or the intuitionistic hedonism of Henry Sidgwick. In both, only one kind of thing is good, namely pleasure. Defining health in terms of such a notion of the good would involve identifying it with pleasure.

[5] This serves in part to overcome the counterintuitive results of identifying all good with, for instance, pleasure.

[6] We are indebted to Brink for these examples, and for his clear presentation of these and other problems of the subjectivist account of the good. See his 1989, chapter 8.

[7] We follow Scheler in our use of the term 'good'; it refers not to a value quality considered in itself and independently of some bearer (for this datum we reserve the term 'value'), but to some *thing* which has value (*Wertdinge*). We do not mean to imply that there are not other meanings of the term 'good' but only to distinguish the two meanings we think central to the analysis of well-being.

[8] While there may be other categories of good that are relational in this sense, we are concerned here only with that category which we call the beneficial.

[9] In Hildebrand's analysis, which we closely follow, the category of beneficial goods is termed the "objective good for the person." Such a good is, in his words, "something which is objectively in my true interest, which has a beneficent character with respect to my person and which is in the direction of my good" (1953, p. 50). The second sense of good, i.e., "my good," is that comprehensive good the actualisation of which we are calling well-being. Cf. Hildebrand, 1980, ch. 6.

[10] Desire is not necessarily motivated by the value of something. One can, e.g., desire something simply because it gives pleasure, or simply because it is beneficial to oneself; in neither case is it the value of the thing as such which motivates. See Hildebrand, 1953, ch. 1.

[11] For a more thorough analysis of the value objectivism we have presented here, see, e.g., Scheler, 1973 and Hildebrand, 1953 and 1980. For a somewhat different analysis that nonetheless has many similarities, see Butchvarov, 1989.

[12] The meaning of the 'state of well-being' refers here to living beings, and is not rightly referred to as a state where 'state' is understood in a static sense.

[13] On the objectivity of conscious states and experiences and the various meanings of subjective and objective, see Hildebrand, 1991, ch. 5; Seifert, 1987, chs. 6 and 7, and his article in this volume.

[14] 'Not feeling well' can have various causes: physical, psychological, emotional, spiritual, moral. When a person experiences, e.g., persistent anxiety, unhappiness, sadness or guilt she is in a real sense 'unwell' even when such states do not have bio-medical causes or cannot be 'cured'.

[15] These two senses of satisfaction are clearly distinct; one could gain the desired goal and yet not feel satisfied or fulfilled.

[16] We do not mean to imply here that well-being is the only or even primary motive for acting, choosing or desiring. On the varieties of motivation, see Scheler, 1973, and Hildebrand, 1953 and 1980.

[17] It should be recalled that in speaking of the human person, we refer not simply to 'a nature' but to the nature of a living being. The phenomenon of life is essential to that of well-being. On the centrality of the phenomena of life for a personalist concept of health, see Seifert in this volume.

[18] See also Wallace, 1978, ch. 1.

[19] Both Kass and Greene formulate the concept of the natural norm in a way consistent with evolutionary change though it need not be so construed.

[20] This point is well expressed in the personalism of Mounier: the "central affirmation" of personalism is "the existence of free and creative persons" and, as such, "it introduces into

the heart of its constructions a principle of unpredictability which excludes any desire for a definitive system" (Mounier, 1952, p. xvi).

REFERENCES

Boorse, C.: 1975, 'On the distinction between disease and illness,' *Philosophy and Public Affairs*, 5(Fall), 49-68.

—— 1977, 'Health as a theoretical concept,' *Philosophy of Science*, 46, 559.

Brandt, R.: 1979, *A Theory of the Good and the Right*, Clarendon, Oxford.

Brink, D.: 1989, *Moral Realism and the Foundations of Ethics*, Cambridge University Press, Cambridge and New York.

Butchvarov, P.: 1989, *Skepticism in Ethics*, Indiana University Press, Bloomington.

Engelhardt, H.T.: 1976, 'Ideology and etiology,' *Journal of Medicine and Philosophy*, 1(3), 256-268.

—— 1981, 'The concepts of health and disease,' in A. Caplan, H. T. Engelhardt, and J. McCartney (eds.), *Concepts of Health and Disease: Interdisciplinary Perspectives*, Addison-Wesley, Reading, Massachusetts, pp. 31-45.

Greene, M.: 1978, 'Individuals and their kind: Aristotelian foundations of biology," in S. Spicker (ed.), *Organism, Medicine, and Metaphysics*, Reidel, Dordrecht, pp. 121-136.

Hildebrand, D.v.: 1953, *Ethics*, Franciscan Herald Press, Chicago.

—— 1980, *Moralia*, Verlag Josef Habbel, Regensburg.

—— 1991, *What is Philosophy ?* Routledge, London and New York.

Kraut, R.: 1979, 'Two conceptions of happiness,' *Philosophical Review*, 88(2), 167-197.

Kass, L.: 1985, *Toward a More Natural Science*, Free Press, New York.

Mounier, E.: 1952, *Personalism*, Notre Dame University Press, Notre Dame and London.

Nordenfeldt, L.: 1987, *On the Nature of Health*, Reidel, Dordrecht.

—— 1993a, 'Concepts of health and their consequences for health care,' *Theoretical Medicine*, 14(4).

—— 1993b, *Quality of Life, Health and Happiness*, Avebury, Aldershot, Singapore, Sydney.

Pörn, I.: 1984, 'An equilibrium model of health,' in L. Nordenfelt and B. I. B. Lindhal (eds.), *Health, Disease, and Causal Explanations in Medicine*, Reidel, Dordrecht, pp. 3-9.

Railton, P.: 1986, 'Facts and values,' *Philosophical Topics*, 14(1), 5-31.

Rosati, C.: 1995, 'Persons, perspectives, and full information accounts of the good,' *Ethics*, 105, 296-325.

Scheler, M.: 1973, *Formalism in Ethics and Non-Formal Ethics of Value*, Northwestern University Press, Evanston.

Seifert, J.: 1987, *Back to Things Themselves*, Routledge & Kegan Paul, New York and London.

Toulmin, S.: 1975, 'Concepts of function and mechanism in medicine and medical science,' in H. T. Englehardt and S. Spicker (eds.), *Evaluation and Explanation in the Biomedical Sciences*, D. Reidel, Dordrecht, pp. 51-66.

Wallace, J.: 1978, *Virtues and Vices*, Cornell University Press, Ithaca and London.

Whitbeck, C.: 1981, 'A theory of health,' in A. Caplan, H. T. Engelhardt, and J. McCartney (eds.), *Concepts of Health and Disease: Interdisciplinary Perspectives*, Addison-Wesley, Reading, Massachusetts, pp. 611-626.

Wright, G.H.v.: 1993, *The Varieties of Goodness*, Thoemmes Press, Bristol, England.

MANUEL LAVADOS

EMPIRICAL AND PHILOSOPHICAL ASPECTS OF A
DEFINITION OF HEALTH AND DISEASE

I. INTRODUCTION

The first part of this article undertakes the task of examining the
phenomenon of disease as it is encountered in the clinical setting, with the
aim of adequately determining the concept. We shall find that disease
constitutes a family of concepts, which are related but distinguishable.
The categories of disease, understood broadly, which shall emerge in our
examination are illness; disease, understood narrowly; and injury. The
narrow determination of 'disease' falls within the extension of the broad
genus of disease in conformity with an unfortunate ambiguity of
language.

Methodologically, my analysis is motivated by a clinical-
phenomenological approach, in which I turn to and describe the
phenomenon of disease as it is encountered in the clinical setting.
Attending to the thing itself (disease) is a necessary prerequisite to
evaluating attempts at explanation. Medicine, after all, is not initially a
theory about the body and how it works so much as it is a practice,
involving complex interactions between doctors and patients, in the
interest of bringing about healing; it is a *techné iatriké*. This conception
of medicine as primarily practical, rather than theoretical, is supported by
the fact that in the most important medical textbooks, there are no
references to a general theory of disease or health. Thus, I must first
describe 'disease' as it occurs in the practical, clinical setting.

Only given this clinical-phenomenological characterization can we
proceed to assess various theories of health and disease. This shall be the
aim of the second part of this article, in which I shall describe and
evaluate two prominent theories of health and disease on the basis of the
conclusions drawn from the first part. The theories of health and disease
which I propose to analyze are the biomedical, and the holistic. I shall
suggest that these two theories, usually thought of as mutually exclusive,
are compatible and emphasize different aspects of the nature of the
human being. The biomedical and holistic accounts of health and disease
will prove complimentary based on our account of the phenomenon of

P. Taboada, K. Fedoryka Cuddeback and P. Donohue-White (eds.), Person, Society and Value:
Towards a Personalist Concept of Health, 187–206.
© 2002 *Kluwer Academic Publishers. Printed in Great Britain.*

disease (understood broadly), which includes the compatible concepts of disease (understood narrowly) and illness, respectively.

II. THE SEARCH FOR A STARTING POINT

As Jacques Maritain has emphasized (1992, p. 931), there is no knowledge without intuition, and this affirmation is especially salient when we try to capture intellectually the most fundamental aspects of reality. Part of the difficulty we find in a theoretical approach to what we call 'health' or 'being healthy' is that it corresponds to a primary and fundamental dimension of living organisms, thereby making it extremely easy to lose track of that about which we are theorizing. For this reason, we must turn to the things themselves, and our intuitions thereof, in order to ground theoretical speculation concerning the nature of health and disease.

In a normal state of being, we experience a spontaneous, relatively permanent, fluctuating and dynamic state of 'well-being' that extends to all the dimensions of our being and living, and includes what we call 'health' or 'being healthy' as one of its components. In this state the human being feels complete and functions as a whole that can cope with the usual requirements of life without pain or discomfort. But the feeling of 'well-being' is also fragile: sadness, anguish, suffering and pain usually accompany our lives. But just as 'health' is a component of 'well-being', that which threatens 'well-being' may threaten 'health' or it may not. For instance, one might feel very sad upon losing one's job; one might feel anguish upon losing a close friend; and one might suffer upon 'stubbing' one's toe. However, while these experiences may detract from one's well-being, possibly drastically, they clearly cannot be said to detract from the state of one's health. For this reason, it seems obvious that not all the experiences usually valued as negative and that alter our feeling of well-being correspond to 'being' or 'feeling' ill (accordingly, none of the above examples can be called 'illnesses' in any meaningful sense of the word).

Given this brief description of health as a component of 'well-being', it seems to me that in relation to health and disease there are two major intuitions. The first is the idea that health refers to a certain aspect of a perfection of living, and in this sense health realizes the idea of a good. Disease, as the converse of health, appears to be an evil since it deprives an organism of a good that it is owed in some way. This seemingly obvious intuition shall be more formally substantiated in what follows. The second intuition is related to the fact that health, in the case of the human being, is a particular good and thus does not correspond to the total good of the individual as is proposed by the World Health Organization which defines health as a state of "complete physical, mental, and social well-being and not merely the absence of disease or infirmity" (WHO, 1958). In light of these intuitions, a difficulty immediately arises: what kind of good is health for living organisms in general and the human being in particular? These are the questions with which we shall concern ourselves in what follows.

III. ILLNESS, DISEASE, AND INJURY

In the English language a cluster of words are used to refer to the realities that concern us here. The most important of these are *disease, illness,* and *injury.* But there are many more: *sickness, disorder, defect, lesion, affliction, dysfunction,* etc. In Spanish, the term *enfermedad* (the English equivalent could be *infirmity*) comes from the Latin *in-firmus* which means etymologically 'lack of firmness' or 'lack of strength' (Coromina, 1976, p. 233). The term *disease* literally means 'lack of ease'. Dis-ease indicates an alteration in 'ease'. The latter term *ease* has a variety of meanings that refer to the absence of work or effort, of physical pain, disturbance, excitement or discomfort; absence of anxiety, anguish, apprehension or something that irritates or disturbs the mind; absence of difficulty or major work. This literal etymology may be interesting and potentially instructive, but, at this juncture, it is unclear whether it is insightful in terms of the way in which we actually use 'disease' language.

As Seguin (1982, p. 191) points out, three universal predicates of the phenomenon of disease are suggested by the observation and experience of peoples: (1) deficiency (*asthenia, infirmitas*); (2) positive damage (*nosos, morbos*); (3) felt or experienced damage (*pathos, aegrotatio, dolentia*). On the basis of these categories, I would like to propose that the terms used to refer to the various states or processes of 'nonhealth' seem to refer to three major categories of phenomena that may be identified in English with the words *disease, illness,* and *injury.* It is now incumbent upon us to examine whether or not this proposition is justifiable based upon the clinical experience of disease (understood broadly, i.e., the converse of health).

The Experience of Illness: A Clinical Point of View

What kind of experience do we call illness? Normally the physician has access to this type of experience when a person – a patient – requests medical attention, and I think this 'search for help' can teach us something about the distinctive aspects of the experience of illness that enable us to differentiate it from so many other experiences that are negatively evaluated.

When a person experiences illness, he or she is justified in seeking medical attention. But this begs the question. When acutely afflicted with severe, unusual, and disabling symptoms, people almost invariably seek medical diagnosis and treatment, and, accordingly, can be said to be ill. However, many people who are ill do not visit doctors. A visit to the doctor is not an automatic response to a symptom; it actually occurs only in a minority of people who consider themselves sick. In a survey during a 2 to 4 week period, about 75% of the general population experienced symptoms they recognized as being due to illness, but only 25% of them consulted doctors and they were not necessarily sicker than those who did not. In fact, the two groups were indistinguishable when the number and type of symptoms were compared (Barsky, 1981, pp. 492-498). However, regardless of whether or not persons seek medical attention in particular cases, the subjective experience of symptoms – usually pain, dysfunction, or disability – supplies the foundation of our concept of illness.

But pain is experienced in e.g. burning oneself, and although one might seek medical attention for a burn, we would not want to describe the experience of being burnt as symptomatic of an illness. The experience of symptoms such as pain seems to be a necessary but insufficient condition for illness. For a more accurate description it seems necessary to add to the experience of illness the fact that the subject accurately perceives that the cause of his or her indisposition is incorporated in his or her being; the patient feels there is something wrong with himself or herself. As noted by Culver and Gert (1982, pp. 63-85), a smoke-filled room can cause labored breathing and extreme cold can cause pain and the loss of ability to move one's limbs easily. Nevertheless, these experiences are not illnesses (nor are these conditions diseases). However, if the previously-mentioned situations affect a person so that he or she continues to suffer the symptoms when the special external situations are no longer present, then that person experiences illness. Thus, a person suffering from an illness must have a condition not sustained by something distinct from him or herself. The expression "distinct from him or herself" does not refer only to external elements; rather, for example, invasive medical instruments are not external but remain distinct from the body of the patient. On the contrary, bacteria or toxic substances are neither external nor distinct elements inasmuch they become biologically incorporated into the organism. The fact that illness is due to some element which is part of the person – or biologically incorporated and indistinct – also determines that it cannot be removed without special skill or equipment, which is one of the reasons the patient seeks (or is at least justified in seeking) medical assistance (Culver and Gert, 1982, p. 73).

The impact of illness – especially when it is severe – not only leads the person to seek assistance due to a wrong which is incorporated into his or her being, but, more fundamentally, it radically transforms an active organism into a passive and vulnerable receptivity wherein the patient can no longer enjoy being the source of action and choice. This observation has been salient to clinicians for quite some time (see, for example, Peabody, 1927). As was noted earlier, one is accustomed to coping with a certain quota of pain and suffering in a fluctuating and dynamic state of health. It is when this pain and suffering becomes noticeable or even overwhelming and is correctly attributed to an indistinct cause that one is ill and justified in seeking assistance. The 'self', in its biological affliction, becomes transformed into something alien, foreign, unfamiliar. The action and freedom enjoyed in health is lost, to a greater or lesser degree, in illness as the pain is either sufficient to limit one's normal activity, or noticeable in that activity. In illness the patient perceives a threat to his or her integrity as a person, insofar as the normal good of free activity to which we are accustomed is lost, to a greater or lesser degree. Some clinical observations of pain (the symptom doctors are most frequently consulted about) and suffering suggest this is so. People in pain frequently report it when they feel the pain to be out of control or overwhelming, when the source of pain is unknown, or when pain is chronic (Cassel, 1982, p. 641). In all these situations people perceive pain as a threat to their continued existence – not merely to their lives, but to their integrity as persons. Although vulnerability is an essential part of the human condition, it is revealed to be more so in illness. Insofar as illness deviates from the normal threshold of pain and suffering which accompanies our healthy lives, and limits or threatens to limit the activities of our routine healthy lives, illness is correctly perceived as a threat to the integrity of our lives.

In general, the term "illness" corresponds to the experience of *having something wrong with oneself* and having something wrong within oneself in this case is to have a condition wherein the person is suffering or has increased probability of suffering some evil whose cause is indistinct from his or her body. The most evident of such evils are pain, disability, and death (Culver and Gert, 1982, p. 63-85). Insofar as these evils are sufficient to warrant the 'search for help', illness represents a fundamental change in the subject, as he or she makes a transition from activity to passivity, freedom to limitation.

Disease as a Principle of Illness

In its customary use, both inside and outside of medical practice, the term *disease* seems to refer to the 'entity' which is the principle of the particular experience of infirmity signified by the word 'illness' (Cambell, 1979, p. 757). Starting with this common sense perception, we can begin by proposing that disease is the principle of illness. Our proposition renders our definition of 'disease' in terms of its relation to our previously explained notion of 'illness'. The critical relation is that of a *principle*. How then are we to make sense of this relation? According to Aristotle (*Metaphysics* V.I), a principle, in its most general meaning, defines that from which something derives its being, becoming, or knowledge. Given this triad of a definition, it is not hard to see that several standard definitions of disease and illness are based upon it: 1) disease as that which explains or causes specific clinical phenomena (that from which illness derives its being); 2) disease as the origin of a process of change (that from which illness derives its becoming); 3) disease as simply that which serves to gnoseologically classify a specific clinical entity (that from which knowledge of illness is derived).

Firstly, the idea of disease as the cause that produces the clinical symptoms is illustrated by the theory proposed by Virchow when he affirms that the origin of all illness is a pathological process that ultimately resides in the cells of the organism, not in its tissues or organs (Virchow, 1860). The idea of disease as a principle of the "becoming" of an illness characterized by the appearance and development of symptoms and signs associated with a particular illness that follow a specific course in time was proposed by Sydenham (1981, pp. 145-155). Finally, the idea of disease as a principle of knowledge is illustrated in modern medicine by the conception of disease as a set of facts or clinical conditions or phenomena that tend to associate themselves statistically, receiving the names of "syndromes" or "disorders" and which are useful to the physician in making empirical diagnostic or therapeutic decisions, even though he or she does not know their cause or physiopathology.

Injury and Disease

The broad notion of disease that we have been explaining, in terms of illness and disease, must be distinguished from the notion of injury. Injury, as positive damage, is, of course, more than a subjective experience. However, it is not the case that injury is another principle of illness, and, accordingly, when one is suffering from, e.g., a broken arm, one is not necessarily ill. The general meaning of injury is "damage inflicted to the body usually by an external force", or more fully, "a disruption of the integrity of a tissue or an organ by external forces that are usually mechanical but also chemical, electrical, thermal or radiant" (Jennings, 1991, p. 976).

As proposed by Jennings (1991, p. 976), injury and disease differ with regard to at least four features. First, injury refers to the impact of a force, not a living process as in disease. Second, injury involves damage at the higher level of tissues and organs not at the cellular level, as we saw was the case with disease on Virchow's account thereof. The third difference lies in the nature of the cause-effect relationship. Consider exposing a living organism to radiant energy. The initial damage is a burn. As with all injuries, the dose-response and dose-effects relationships are graded and quantitative and have clinical thresholds. The damage is not self-propagating, and it is followed by healing. The other consequence of this exposure is a disease process: increased frequency of genetic mutation (response) and consequent cancer (effect). In disease, there is no dose-effect relationship: a cancer is not worse because it has been induced by more radiation. The fourth distinction is histological: cellular changes at the site of the injury are the tissue damage itself or a normal reparative response. In contrast, with disease there is an uncontrolled pathological cellular process that does not tend to heal; on the contrary, it is destructive.

To sum up, the above considerations, stemming from our thorough-going clinical phenomenology, suggest that for an accurate and foundationally-sound philosophical reflection about health and disease, it is necessary to acknowledge the specificity of and relationship between these three kinds of phenomena: illness as a particular subjective experience, disease as a principle of illness, and injury as physical damage to the living organism.

Of these three phenomena, two of them – disease and illness – provide the basis for two different ways of conceptualizing disease, understood broadly, and, conversely, health. In the first case, disease can be understood as something strictly cellular which is the object of the biomedical sciences. This first conception, usually called the biomedical conception of disease and health, is based upon a narrow concept of disease and attempts to understand disease by considering the human person as a member of an animal species, and thus conceives of disease as a universal phenomenon that affects all living beings, all of whom are subject to various cellular anomalies. Within this model, disease or health are objective phenomena subject to observation and empirical analysis. In the second case, disease is understood as a uniquely human affliction, and, accordingly, all the aspects we have included in our description of illness are emphasized. This second conception of disease, usually called 'holistic', understands disease in terms of the subjective experience thereof and rejects the former strictly theoretical (objective) account.

IV. THE BIOMEDICAL MODEL OF HEALTH AND DISEASE

A clear biomedically-oriented theory has been presented by Christopher Boorse in his classic article "On the Distinction between Disease and Illness". One of Boorse's main objectives is to develop definitions of health and disease that are purely descriptive (objective) thereby excluding any evaluative (subjective) elements. In terms of our preceding clinical analysis, Boorse's project involves proposing a distinction between health and disease which somehow ignores the phenomenon of illness, of which disease is the principle. Boorse is solely concerned with the theoretical notion of disease and disregards the clinical contexts in which such notions might be employed (Boorse, 1975). He calls *normativist* or subjective those theories that affirm that it is impossible to define health or disease without incorporating value-judgments or normative conceptions. The normativism Boorse has in mind claims that to call a condition unhealthy signifies, at least in part, a condemnation of it; to call a condition health signifies its desirability for an individual or society.

In practice, it is apparent that unbridled normative concepts of health and disease do not sufficiently reflect certain elemental facts and prove counter-intuitive. If undesirability defines a condition as unhealthy, it would follow, for example, that the lack of the manual dexterity of a great pianist could be a "disease" for an individual who does not have it but wants it, which is obviously absurd. Likewise, if desirability defines a condition as healthy, the child who luckily evades his exam due to an incidence of influenza must be regarded as healthy. These examples are absurd insofar as the presence or absence of so-called 'diseases' would be contingent upon very specific states of mind or goals which may differ among persons and intra-personally over time. There must be some standard for deciding whether or not a given condition is to be generally regarded as a disease. Insofar as the counter-examples provided above are somewhat anomalous, maybe the answer lies in assessing the general desirability or undesirability of a given state. As Boorse points out, it is hard to determine whether conditions like infertility are desirable or undesirable, and possibly those who consider it to be *prima facie* undesirable assume that most people want to be able to have children. But the consequence of this position would be that writers of medical texts would have to make an empirical survey of human preferences in order to determine whether a condition (e.g., infertility) is a disease (Boorse, 1975, p. 66), which does not seem to signify great progress for the normativist position.

From our critical analysis of normativism, we can conclude that the distinction between desirability and undesirability alone cannot maintain the distinction between health and disease in any meaningful way. This rudimentary normativism is simply too amorphous, insofar as undesirability is neither necessary nor sufficient to render a given state as disease. However, we can extract at least the following idea from our discussion of normativism: whenever an individual – the patient or the doctor – classifies a condition as unhealthy, the condition is being judged in comparison to an ideal norm of some sort. The structure of health judgments involves a comparison with an ideal and through such comparisons the normal is established by distinguishing it from what is abnormal. However, this begs the question as to how we are to maintain the distinction between this ideal norm and that which significantly deviates from it. Health and disease are admittedly normative judgments, but they are not grounded in personal proclivities. The distinction must be maintained objectively.

Statistical Normality

A first objective criterion by which a condition could be defined as healthy or unhealthy would be statistical normality. If this criterion were adequate, then all conditions that do not deviate significantly from the statistical average would be considered healthy, and all conditions which do deviate significantly would be considered unhealthy. Statistical normality is the Boorsian solution to the problem of the objective criterion of health, and one can be said to be healthy, according to Boorse (1997), if and only if, one maintains species-typical functionality consistent with one's reference class within a statistically normal range.

Two examples will serve to show that this criterion is inadequate. Firstly, many deviations from the average are not unhealthy. For instance, the unusual strength of an Olympic weightlifter is significantly deviant from the statistically normal range of strength within his or her reference class. The statistical distribution of a given reference class will tend to conform to a bell-curve, with deviations from the norm occurring at the extremes of the relevant axis. However, any determination of which abnormalities are pathological and which are positive appears to import a more subjective, normative judgment. Secondly, there may be reasonable doubts about whether certain statistically "normal" conditions correspond to healthy conditions. According to Boorse, menopause and senile dementia among the aging cannot be regarded as diseases insofar as their occurrence is statistically normal within the relevant reference classes. However, diseases such as osteoporosis and arthrosis were also once regarded as normal due to their perceived ubiquity and irreversibility among the elderly. It appears dubious that statistical normality can be equated with health in the case of the elderly.

Clinical Normality

If the criterion of statistical normality is inadequate as an objective means of distinguishing health from disease, what other kind of objective norm could be more appropriate? It seems to me that the project of devising a wholly objective norm for making disease judgments is simply wrongheaded. The fallacy of theories such as Boorse's is that they attempt to divorce the scientific concept of disease from that of the clinical notion of illness. Recalling our earlier reflections on illness and disease, if disease is simply the principle of illness (in three senses), then to consider the former while disregarding the latter is quite philosophically dangerous. However, our suggestion that any consideration of disease cannot proceed oblivious to the clinical notion of illness of which it is the principle, certainly does not commit us to the naïve variety of normativism criticized in the previous section, as well as by Boorse.

Therefore, let us return to the clinical experience of illness in the hope that it will shed some light on the norm employed in making the distinction between health and disease. When the physician judges a condition to be unhealthy, or more precisely pathological, he does not do so based primarily on a possible deviation or lack of deviation from statistical averages; this much is clear. Rather, the clinician establishes a correlation between what he observes – a certain defect for example – and the foreseen impact or influence of that defect on the individual's overall functionality. In establishing this correlation, the patient's subjective experience is related to a pathological disease condition, as the principle of the illness. Recall that the disease is not merely an ontological concept, but an epistemological concept which illuminates the potential 'becoming' or the natural history of the disease. Thus, from the subject's descriptions and the physician's scientific observations of symptoms, the physician is able to foresee the impact of the disease upon the patient. For example, the neurologist judges that memory defects are "pathological" if he observes or predicts that the individual can not or will not be able to function adequately in his social, family or work activities as a result of memory defect.

Another example that may serve to illustrate our suggestion is found in clinical judgments concerning the existence or absence of heart disease. The expert cardiologist Eugen Braunwald writes:

A cardinal principle useful in the evaluation of the patient with suspected heart disease is that the myocardial or coronary function which may be adequate at rest may be inadequate during exertion. Thus, a history of chest discomfort and/or dyspnea which appears only during activity is characteristic of heart disease, while the opposite pattern, i.e., the appearance of these symptoms at rest and their remission during exertion, is rarely observed in patients with organic heart disease (Braunwald, 1987, p. 863).

The same idea is emphasized in the definition of mental disorders. In the American Psychiatric Association's Diagnostic and Statistical Manual of Mental Disorders (DSM-IV), each mental disorder is conceptualized as a clinically significant behavioral or psychological syndrome or pattern. For every diagnosis in DSM-IV, the symptoms by which the person meets the criteria threshold must cause

clinically significant distress (a painful symptom) or impairment (disability) in social, occupational, or other important areas of functioning. In addition, this syndrome or pattern must not be merely an expected and culturally sanctioned response to a particular event (e.g., the death of a loved one). Whatever its original cause, it must currently be considered a manifestation of behavioral, psychological, or biological dysfunction in the individual (DSM-IV, 1994, pp. xxi-xxii).

Specifically, this is the way that DSM-IV has clinically defined dementia (DSM-IV, 1994). The essential feature of dementia is the development of multiple cognitive deficits that include memory impairment and at least one of the following cognitive disturbances: aphasia, apraxia, agnosia, or a disturbance in executive functioning. But the "cognitive deficit must be sufficiently severe to cause impairment in occupational or social functioning and must represent a decline from a previously higher level of functioning" (DSM-IV, 1994, pp. 134-35).

Clearly, this does not mean that there are no pathological conditions characterized by an absence of symptoms, or where the functional repercussions are very slight or even non-existent. Nonetheless, when the physician evaluates a condition, he knows because of follow-up studies or by simple prediction that the defect he has observed places the individual at risk of developing difficulty in functioning. Therefore, a disease judgment may be in order despite the absence (present or future) of the subjective experience thereof.

The clinical judgment that establishes what is healthy by distinguishing what is not, proceeds as if its 'norm of reference' were simply the fact that a specific individual may or may not function as he should function according to the expectations for the biological species that characterizes him as an organism of a specific type.

If this is correct, then it is possible to suggest that the distinction between healthy and pathological or diseased refers to the nonexistence or existence of a defect in structure and/or function that prevents the individual – in variable degrees – from exercising the functions or activities that characterize the individual as an organism of a specific type. It is important to note that this account does not claim to be 'so objective' that if cannot take cognizance of illness as the ordinary impetus for the disease judgments of physicians. However, it remains strongly objective in the sense that the distinction is based upon functionality in terms of species design, and not subjective expectations. For instance, whether or not one's hands are adequate for attaining the status of a piano virtuoso is irrelevant to one's capacity to function according to one's nature. And this account of the biomedical model of health and disease does presuppose the philosophical notion of nature – that is the idea that a living organism is structured according to an end. This end is a state of activity or functioning that is intrinsically good for the organism (i.e., not subject to individual desires or expectations).

The Living Organism and the Notion of Nature

Now let us try to place this clinical idea about the distinction between healthy and unhealthy according to the biomedical model within the framework of a more general theory about the nature of living organisms, including the human being.

We must initiate this discussion with the basic question of what it means to have a nature. Once again, let us seek guidance from Aristotle. In his *Physics,* Aristotle makes the case that for something to have a nature is for it to have "within itself a principle of motion and of stationariness" (Physics II, 129 b15). This principle is the *telos* of the entity in question, and it seems reasonable to suggest that one of the essential features of biological organisms is that they are structured according to and operate with reference to some end. It may be possible, then, that nature, or function unto an appropriate end, could supply the objective norm which is employed in clinical judgments. This suggestion has, in fact, been offered by King: "The root idea of this account is that the normal is the natural. The state of an organism is theoretically healthy, i.e., free of disease, insofar as its mode of functioning conforms to the natural design of that kind of organism" (King, 1945, pp. 493-494).

It is now incumbent upon us to decipher the end toward which living organisms are directed according to their natures. Organisms are vast assemblages or systems and subsystems which, in most members of a species, work together harmoniously to achieve a hierarchy of goals or ends. An organism and its parts are directed toward specific goals. Based on this conception of goals, cells are goal-directed toward metabolism, the heart is goal-directed toward supplying the rest of the body with blood; the nervous system is goal-directed toward the analysis of information which permits the organism to develop behavior adapted to a complex, changing environment, etc. Clearly, cells, the heart, and the nervous system are critical parts of the organism with accordingly critical goals, but what are the ultimate goals of the organism, to which these parts contribute? We suggest that the contributions and interrelationship of these partial (or specific) functions are directed toward the accomplishment of the greater biological ends that are common to all species, namely, survival and reproduction. It is for the sake of these two ultimate ends that the organism needs to be able to develop the activities characteristic of the particular kind of organism to which the species belongs. Despite the rather divergent notion of normality that we observed above, Boorse also reaches the conclusion that "medical functional normality was presumably relative to the goals physiologists seem to assume, viz., individual survival and reproduction" (Boorse, 1997, p. 9).

An organism is healthy, then, only insofar as it 'functions in harmony with its nature', that is, only insofar as each organ, system and subsystem fulfills the purpose or function for which it exists, which is ultimately the goals of survival and reproduction. These two ultimate goals are the *telos* of all other systemic functions, which are constitutive of and contribute to the organism.

Insofar as our account of normality is determined according to function, it is teleological. However, while "the idea of teleology is normative," it is not subjective (Jacobs, 1986, p. 392). Organic teleology, unlike intentional teleology, for instance, is "altogether amenable to empirical and nomological specification", such that "the good of an organism or the well-functioning of a part of it can be objectively explained with reference to what we know to be requirements for survival [and] reproduction" (Jacobs, 1986, p. 392). I must agree with Woodfield's view that an end of a biological organism is essentially a state or activity that is intrinsically good for this kind of organism (Woodfield, 1976).

If health corresponds to functioning in conformity with the natural design of a given kind of organism, disease may be defined as a state wherein the organism is unable to exercise one or more of the functions typically performed by members of the species, where such deviation from the norm significantly hampers one's ultimate goals of survival and/or reproduction. Now, since the purpose exists for the good of the organism, it may be suggested that since disease corresponds to a certain state of deprivation of proper functioning, it realizes the idea of evil, namely the deprivation of, not merely *some* good, but a *due* good. For beings of the human species, not being able to fly is not an evil because that capacity is not part of the functional organization that is contributive to and constitutive to the ultimate good of our species. For a living organism that develops progressively, the determination of health must also include reference to the temporal dimension; the fact that a certain organism does not possess language, for example, is not evil unless it does not possess it in the conditions and at the time that it should possess it.

The preceding has confirmed our initial intuition that health is a component of the good of an organism, and that disease, conversely, is an evil. Further, it has explained what kind of good health is, relative to our notion of natural functioning.

Natural Functioning and the Definition of Disease

It seems reasonable to me to suggest that any adequate definition of disease ought be able to do the following: 1) explain the norm of reference used to distinguish health from lack of health; and 2) outline the features or attributes which are common to disease in every living organism (including the human being). From what has been proposed, I would like to conclude the following: 1) the norm of reference according to which a certain condition may be classified as healthy or unhealthy is the degree to which biological functioning conforms to the organism's natural design; 2) Function that is not in harmony with nature is the only attribute common to all diseases in whatever kind of living being (vegetable or animal). The objective norm of natural function which has been proposed underscores the minimum requirement for distinguishing between health and lack of health across the species. It remains to be seen

whether or not the phenomenon of disease might involve something more
in the unique case of the human organism.

V. THE HOLISTIC MODEL OF HEALTH AND DISEASE

In fact, the holistic model of health and disease underscores that which is
unique to the human organism. According to the preceding clinical
phenomenology, holistic models stress the subjective aspect of human
disease that was described with the term 'illness'. In contradistinction to
this, biomedical models attempt to define health and disease objectively
according to our understanding of disease, understood narrowly, based on
the way in which living organisms are structured and function. As was
noted above, whether or not an organism functions according to its nature
is a norm applicable to all varieties of organisms (vegetable and animal).

Although in the literature the biomedical and holistic models are
generally presented as mutually exclusive, they share some common
ideas and the philosophical notion of nature proposed in the previous
section can allow us to build a conceptual bridge between them.
Ultimately, we shall find that the holistic model can supplement the
biomedical model for the human organism, and deepen our understanding
of the kind of good which health is in this very unique case. In this
context, we shall briefly summarize some of the basic conceptual
elements of the holistic model.

As in the biomedical model, the ideas of functionality and purpose
repeat themselves as key elements. The difference lies in the fact that
these ideas do not refer to the human being as a member of a particular
animal species, but rather, they refer to the fact that the human person is a
free being who determines for him or herself the purposes that organize
his or her life. The *telos* involved in the holistic model involves
intentionality, rather than simply the organic *telos* of the biomedical
model. This opens the door to a more subjective normativity, of the
variety disparaged by Boorse. The starting point of the holistic theory of
health is to view the human being as an active creature living in a
network of social relationships. Health and illness are therefore viewed as
phenomena that involve the individual as a social agent (Nordenfelt,
1993).

The holistic theory of health emphasizes two cardinal features: first, a
particular type of feeling, a feeling of well-being in the case of health and

a feeling of suffering in the case of illness; and second, the individual's ability to perform the actions that he or she needs or wants to perform in daily life. The first feature is linked very closely to one of the essential features of our account of illness, namely, the subjective experience of symptoms. The second feature is also linked to our account of illness in terms of the impact of illness being a transition from ordinary, everyday activity and freedom to passivity and limitation.

As noted by Nordenfelt, of these two main concepts, it is the second, that of the "ability" to freely perform one's activities, that has the more central place. The standard idea "is to relate ability to a certain class of the individual's goals" (Nordenfelt, 1993, p. 280). What kind of goals? We have already suggested that the holistic model offers a more subjective rendering of 'goals' than the biomedical, and Ingmar Pörn, for instance, agrees with this assertion in suggesting that the goals could be any set of the individual's goals in life (Pörn, 1993). Nordenfelt, however, proposes a critical restriction: "What a human being needs to be able to do in order to be healthy is, in my opinion, to realize 'all' of his or her vital goals" (Nordenfelt, 1993, p. 280). On the basis of this restriction, he defines health in the following way:

> P is completely healthy, if and only if P has the ability, given standard circumstances, to realize 'all' of his or her vital goals, P is unhealthy (or ill) to some degree, if and only if P, given standard circumstances, can not realize all of his or her vital goals or can only partly realize them (Nordenfelt, 1993).

In an effort to explain more precisely the meaning of what Nordenfelt calls "vital goals" it is important to point out that he refers to those goals which all human being must share as a *sine qua non* of whatever particular goals they may individually pursue. An individual's vital goals in Nordenfelt's theory are those whose satisfaction is essential for the individual concerned to achieve a minimum degree of real long-term happiness. However, as Fulford has noted (Fulford, 1993), of the vital goals proposed by Nordenfelt, only survival and reproduction are really essential (i.e., *sine qua non*). The gradual circumscription of goals from Pörn to Nordenfelt to Fulford effectively limits the subjectivity of health and disease, the perils of which have been addressed in our section on the biomedical model.

What I would like to emphasize is that even this kind of holistic theory of health and disease seems to imply the idea that the human being is a "natural being" in the sense that 1) the human being is intrinsically directed toward a certain final state common to all individuals of the species (happiness) and 2) that certain objective conditions are necessary for achieving this end. According to holistic theory, health appears to be an instrumental good that allows an individual to carry out the objectives he or she has freely chosen. Nevertheless, one has to acknowledge that this freedom is objectively limited by conditions dependent on the objective structure (nature) of the human being. If this is the case, then an objective difference between health and disease can be established in terms of the common end proper to given species and the means that are necessary for achieving this end. Thus, health may be conceived as an instrumental good for a being of a given nature, a good that allows this being to achieve the specific ends inscribed in its very nature.

In conclusion, I would suggest the following: 1) as explicitly recognized by biomedical theories and implicitly recognized by holistic theories (at least in Nordenfelt's case), the idea of health seems to correspond to a state of functioning in harmony with nature that places the organism in an intrinsically good state; 2) the intrinsically good state of the organism may be objectively described by referring to the purposes (ends) for which that organism is empirically structured; 3) in the case of organisms that are not free, health seems to be identical with the organisms' total good; 4) in the case of the human being, health cannot be identified with the person's total good. Human health corresponds to a state of functioning in harmony with nature, but only with regard to those functions – corporal and psychic – that are directed toward a pre-determinate set of goals. Thus, human health represents a state of perfection of the nature limited to the dimensions of the person that do not depend on the exercise of freedom. And since the total good of the person includes the exercise of freedom, health can be regarded as an instrumental good with regard to the person's total good, as a state that facilitates the exercise of freedom.

VI. CONCLUSION

Disease imposes limitations on the free construction of the individual's personal being. Illness, like so many other evils our being is exposed to, limits the freedom that Jacques Maritain called "freedom of spontaneity": the spontaneity of every being to act, without external restrictions, according to the being's nature. This freedom of spontaneity which is carried out in varying degrees in the different types of beings in nature, reaches its perfection in the human being, "who being able to go beyond the moment of feelings and to know being and intelligible natures, knows what he is doing, and the purposes of his acts as such; deciding himself, by his own intellectual activity, the purposes of his actions" (Maritain, 1967, p. 181). If disease is what I have proposed, then the ultimate nobility of medicine does not lie in the act of alleviating the pain or suffering that illness brings; its ultimate nobility resides in its cooperation with the exercise of the person's freedom.

Pontificia Universidad Católica de Chile
Santiago, Chile

NOTE

The author acknowledges his gratitude to Jeremy Dorsett and Paulina Taboada for their valuable help in shaping the final version of this contribution.

BIBLIOGRAPHY

American Psychiatric Association: 1994, *Diagnostic and Statistical Manual of Mental Disorders*, American Psychiatric Association Press, Washington, D.C.

Aristotle: 1941, *Physics*, R. McKeon (ed), Random House, New York.

—— 1982, *De L'Ame*, J. Tricot (trans.), J. Vrin, Paris.

Barsky, A.: 1981, 'Hidden reasons why some patients visit doctors,' *Annals of Internal Medicine* 94, 492-498.

Braunwald, E.: 1991, 'Disorders of the heart,' in J.D. Wilson, E. Braunwald, K.J. Isselbacher, R.G. Petersdorf, J.B. Martin, A.S. Fauci, R.K. Root (eds.), *Harrison's Principles of Internal Medicine*, International Edition, McGraw-Hill, Inc.

Boorse, C.: 1975, 'On the distinction between disease and illness,' *Philosophy and Public Affairs* 5, 49-68.

—— 1997, 'A rebuttal on health,' in James M. Humber and Robert F. Almeder (eds.), *What is Disease?* Humana Press, Totowa, 3-134.

Cambell, E.J.M., et al.: 1979, 'The concept of disease,' *British Medical Journal*, 2, 757-62.

Cassel, E.J.: 1982, 'The nature of suffering and the goals of medicine,' *The New England Journal of Medicine* 306, 639-645.

Corominas, J.: 1976, *Breve Diccionario Etimológico de la Lengua Castellana*, Editorial Gredos, Madrid.

Culver, C.M and Gert, B.: 1982, *Philosophy in Medicine, Conceptual and Ethical Issues in Medicine and Psychiatry*, Oxford University Press, New York, Oxford.

Fulford, K.W.M.: 1993, 'Praxis makes perfect: Illness as a bridge between biological concepts of disease and social conceptions of health,' *Theoretical Medicine*, 14, 305-320.

Jacobs, J.: 1986, 'Teleology and reduction in biology,' *Biology and Philosophy* 1, 389-399.

Jennings, D. et al.: 1991, 'Essential hypertension: A sign in search of a disease,' *Journal of Canadian Medical Association* 144, 973-979.

King, C.D.: 1945, 'The meaning of the normal,' *Yale Journal of Biology and Medicine* 17, 493-494.

Maritain, J.: 1992, 'Pas de savoir sans intuitivitè,' in *Approches Sans Entraves Oeuvres Complètes Editions*, Universitaires Fribourg- Editions Saint Paul, Fribourg, Paris, 931-994.

—— 1967, *De Bergson a Sto. Tomás de Aquino*, G. Moteau de Buedo (trans.), Club de Lectores, Buenos Aires.

Nordenfelt, L.: 1993, 'Concepts of health and their consequences for health care,' *Theoretical Medicine* 14, 277-285.

Peabody, F.W.: 1927, 'The care of the patient,' *Journal of the American Medical Association* 88, 877-82.

Pörn, I.: 1993, 'Health and Adaptedness,' *Theoretical Medicine* 14, 295-303.

Seguin, C.A.: 1982, *La Enfermedad, El Enfermo, y El Médico*, Ediciones Pirámide, S.A., Madrid.

Sydenham, T.: 1967, 'Observationes medicae,' in A.L. Caplan, H.T. Engelhardt, and P. McCartney (eds.), *Concepts of Health and Disease: Interdisciplinary Perspectives*, Addison-Wesley, Reading, Massachusetts, 145-155.

Van der Steen, W.J. and Thung, P.J.: 1988, *Faces of Medicine: A Philosophical Study*, Kluwer Academic Publishers, Dordrecht.

Virchow, R.L.D.: 1860, *Cellular Pathology as Based Upon Physiological and Pathological Histology*, Churchill, London.

Woodfield, A.: 1976, *Teleology*, Cambridge University Press, Cambridge.

World Health Organization: 1958, *The First Ten Years of the World Health Organization*, Preamble to the Constitution of the World Health Organization.

SECTION THREE

HEALTH AND SOCIETY

PIET VAN SPIJK

POSITIVE AND NEGATIVE ASPECTS OF THE WHO
DEFINITION OF HEALTH, AND THEIR IMPLICATIONS
FOR A NEW CONCEPT OF HEALTH IN THE FUTURE

> "Defining general terms is not an abstract exercise but a way of shaping
> the world metaphysically and structuring the world politically."
> Daniel Callahan (1973, p. 74)

I. INTRODUCTION

Working as a doctor in an outpatient psychiatric hospital, I had just
finished the treatment of a young homosexual woman who had recently
separated from her girlfriend. She experienced a strong grief reaction and
came into treatment for this reason. She improved rather quickly, and it
was at that moment that I was confronted with the following practical
question: Does a homosexual person, who feels reasonably well, require
further treatment? Is such a person healthy or not? What are the grounds
for a "yes" or "no" in response to this question? The medical library of
the University Hospital where I worked, and the medical literature in
general, was of little help in answering such questions. I was surprised to
find nothing there.

The Technical Preparatory Committee of the WHO must have faced
the same problem of finding adequate literature when they created, under
what seems to have been a certain time pressure, a new definition of
health. At their meeting in Paris, the problem of defining health was
discussed on the 22nd of March, 1946. On that date, several similar
definitions were proposed by a subcommission. A first proposal was the
following: "Health is not only the absence of infirmity and disease, but a
state of well-being and integrity, both physical as well as mental,
resulting from the intervention of positive factors, such as diet, housing
and adequate education" [1] (WHO, 1946a).

A second proposal was: "Health is not merely the absence of infirmity
and disease but a state of physical and mental equilibrium and social
well-being"[2] (WHO, 1946a). Here we find the 2000 year-old idea of
equilibrium. Only about three months later, at the International Health
Conference held in New York, the definitive formulation was found:

*P. Taboada, K. Fedoryka Cuddeback and P. Donohue-White (eds.), Person, Society and Value:
Towards a Personalist Concept of Health,* 209–227.

Health is a state of complete physical, mental and social well-being and not merely the absence of disease or infirmity.

The enjoyment of the highest attainable standard of health is one of the fundamental rights of every human being without distinction of race, religion, political belief, economic or social condition. The health of all peoples is fundamental to the attainment of peace and security (WHO, 1946c).

At the conference, representatives of 61 nations signed the Constitution of the WHO. The animating spirit behind the formation of the WHO was the conviction that the improvement of world health would make a contribution to world peace (Callahan, 1973, p. 78).

In parts II and III.1, I will present a summary of the positive aspects and deficiencies of the WHO definition of health. The largest part of this presentation is to be found scattered throughout the work of a large number of authors. To my knowledge, however, a consideration of the formal and methodological deficiencies of the WHO definition has yet to be undertaken. In part III.2 I will investigate these points more thoroughly. In the course of this, there will emerge a concept which should help in the search for a coherent and practically applicable definition of health. I consciously do not attempt to formulate a better definition; I am concerned rather with sketching out the difficult and complex path toward a good definition, and giving an indication of the wide, fascinating, and challenging philosophical landscape which begins to unfold once the question about the nature of health is asked. To fasten upon a theoretical and practical definition of health before this landscape in its many dimensions has been scientifically and philosophically explored and investigated is something of which one should, in principle, be wary.

II. POSITIVE ASPECTS OF THE WHO DEFINITION

In the following, I would like to present briefly what to my mind are the positive aspects of this definition of health.

1) The WHO definition turns the focus away from the absence of disease and places greater interest in a positive definition of health. For at least the previous hundred years, this had not been the case. People were so busy fighting bacteria, performing major surgery or fighting hysteria, schizophrenia and many other serious diseases, that they nearly forgot

that victory over a disease is not necessarily the same as the recovery of health.

2) For the first time in history, a worldwide organization became interested in the subject of health, which means that now, a lot of people will think, know and talk about this subject. In the past, there had always been individuals who wrote and thought about health. Since 1946, however, it has become a public affair. It was at this moment that a public commitment to health for the whole of mankind was founded -- an event comparable with the worldwide commitment for peace expressed in the creation of the League of Nations in 1920. Since 1946, interest in the problem of health has grown very much, literature on this subject has virtually exploded, and now, this problem has become one of the important topics in our society. All this the WHO initiated in 1946.

The members of the preparatory conference in Paris were not only aware of the importance of what they did when founding the new organization, they also realized that its name was of great importance. There was a long and intensive discussion on the question of the name. At that time, greatest emphasis was placed on the word "world." But the main innovation was certainly the introduction of the word "health" into the organization's name. Instead of calling it "WHO," it would have been easy to call it "WDCO" for "World Disease Control Organization." With the word "health" in its name, one of the first duties of the WHO was to define its own name, which means defining health. And this is what has been done indeed.

3) A third advantage of the 1946 definition of health is, paradoxically, its poor quality. Some of its insufficiencies are so evident that nearly everybody feels inclined to contradict it and try to find a better alternative. This may have been the starting point for many to study this problem more seriously, or even to enter into a long-term engagement in theoretical or practical studies of health.

It is interesting to note that in 1986 the WHO created a new definition of much better quality during the first International Conference for Health Promotion, held in Ottawa. But even though the Ottawa Charter, as it is known, contains a more adequate definition of health, almost no one knows it and its importance is small.[3]

III. NEGATIVE ASPECTS OF THE WHO DEFINITION

There are several criticisms of the WHO definition; a first category concerns its content and the second the methodology underlying its creation. In the following, I will first enumerate several points that focus on the content of the definition. Second, I will focus on methodological points. The problem of constructing a definition of an object or concept as complex as that of health was, it appears, underestimated by the WHO. The concept of health, like concepts such as freedom or love, is rich and variegated. The definition of such concepts must formally and methodologically correspond to this differentiation and richness.

A. *Criticisms of the content of the WHO definition*

1) In the definition, one unknown word, namely "health," is defined with another one that is equally vague. Or as H.T. Engelhardt puts it, "Such a definition of health packs the ambiguity of the concept of health into the ambiguity of a concept of well-being" (Engelhardt, 1981, p. 32).

2) A state of complete physical, mental and social well-being covers, of course, much more than the word "health" reasonably can cover. Complete social well-being, for instance, has to do with having enough money, having the right to vote, and the equal treatment of women, children, foreigners and minorities. It also has to do with good schools and good education. The same can be said about the idea of complete physical well-being. The search for such a state and the rejection of bodily or physical suffering is one of the causes for drug and medication abuse, and one of the reasons for the cost explosion in our health care systems.

One could argue that the WHO definition is a modern rebirth of the age-old myth of what has been called the "Golden Age," "paradise," or the "Kingdom of Heaven." There is nothing wrong with creating such a myth, but it is important to be aware of the fact that it is a myth or a symbol and cannot be understood as a concrete reality. The fact that the WHO also said that the goal of its own activity is health for everybody by the year 2000 shows that it itself did not notice that it was treating a myth as if it were a reality.

3) This vagueness of the WHO definition leads to significant negative consequences for society as a whole. In a world in which politicians have failed to give us complete well-being, the WHO suggests that the

responsibility for all human ills be taken over by medical technicians (Callahan, 1973, p. 82). Therefore, the question of whether someone can make a long ocean journey, enroll in a diving course, of whether insurance should cover the costs of attending a fitness center, etc., are all questions which to a large extent start to depend on the professional opinion of physicians. Likewise, nowadays it is impossible to resolve criminal law cases without finding that medical opinion plays a decisive role in the decision. Whether an eighty-year old patient should receive a kidney transplant, or whether he should be re-animated in the case of heart failure, etc., are all examples of cases in which the physician freely assumes responsibility for his decisions and their consequences. This power which a physician exercises over his fellow man, however, is slowly beginning to show evidence of a boomerang effect: physicians ('demi-gods in white') in this position find themselves placed under excessive demands; they can neither fulfill society's hopes for comprehensive medical care, nor respond decisively to all the important questions of life. In the end, the scorn of a disillusioned public falls upon the physicians, who themselves, with regard to the ultimate things, are as impotent as the rest of society. In the final analysis, we find ourselves with frustrated and over-burdened physicians practicing in a society of individuals wholly disenfranchised in the sphere of health and issues of life.

This current trend places physicians in a losing position (which they unfortunately too willingly accept) and gives society illusory hopes. In addition, if one inflates the term 'health' to such a degree that it becomes the opposite of "all human disorders - war, crime social unrest, ... then one turns health into a normative concept, that which human beings must and ought to have if they are to live in peace with themselves and others... 'Health' can and must be imposed. Here can be no room for the luxury of freedom when so much is at stake" (Callahan, 1973, pp. 82-83).

4) Using words like "complete" or "absolute" in the context of social, medical or biological states is highly problematic. Galen explained that there is an absolute equilibrium and a relative equilibrium. If we take the first as a standard for health, the term becomes useless because no real living human being will ever live in a state of absolute equilibrium or a state of absolute well-being. Therefore, nobody could be called healthy, and everybody must be considered sick. Galen continues by saying that real health is always changing a little, and that it also changes with time, because the health of a child is not the same thing as the health of an old

person. On the issue of where in practical life the boundary between health and illness is to be drawn, he gives as a practicing physician the following rule of thumb:

> that condition in which we do not suffer pain, and are not impeded in the activities of life, we call health; and if anyone wishes to call it by any other name, he will accomplish nothing more by this than those who call life perpetual suffering .... They say that the seeds of all diseases are in us. But they themselves confess these are so small that they escape our perception. Granted, if they wish, that there is in us a certain morbid diathesis, yet so slight and imperceptible as not to harm the possessors .... The unimpaired capacity of function determines health (Galen, 1951, p. 15).

The Swiss philosopher J. Hersch stresses another point in this context. She says that human beings are aware of the fact that they will die. This frightening knowledge of our own agony and death to come, and the uncertainty of what comes afterward, causes within us a continuous suffering. This suffering is inevitably linked to our lives in this world. So, in some way, we all are a little sick all the time, at least sick in knowing about our own mortality. Hersch then says, "Living is knowing how to be sick, it is the same as being a little bit sick most of the time. And in good health, therefore, is he who can bear the prospect of his own death and the probability of being sick" (1981, pp. 1531 ff. - author's translation).

5) The fifth criticism concerns the idea that health is a fundamental right of every human being. L. Kass notes:

> It no more makes sense to claim a right to health than a right to wisdom or courage. These excellences of soul and body require natural gift, attention, effort and discipline on the part of each person who desires them. To make my health someone else's duty, is not only unfair; it imposes a duty impossible to fulfill .... The theory of a right to health flies in the face of good sense, serves to undermine personal responsibility and, in addition places obligation where it can not help and be fulfillable (1985, p. 183).

It should be noted in this context that there are other aspects, often associated with the concept of health without any further question, but which actually are in need of closer consideration. In particular, we should consider the assumption that a) health is something unconditionally good (as is implicitly assumed in the WHO definition), b) health is the most important part of life in general, and c) it is every person's duty to maintain his health, or in cases of illness, to recover their health. I would like to make a few remarks regarding each of these points.

a) The idea that health is something unconditionally good has been questioned by some authors. Meister Eckart in the late Middle Ages said that the search for health distracts men from their search for unity with God. And in the early Middle Ages, several early Fathers saw health as something pernicious (*sanitas perniciosa*) to men, because it is a sign of the absence of God, "a dangerous state, because it easily arouses a feeling of security, arrogance and indifference toward essential needs of the soul and its salvation" (Schadewaldt, 1965, p. 126, author's translation).

b) How important is health for a fulfilled life? Let us take the example of a woman recently telling me that she was just about to get divorced, had serious quarrels with her husband, was reaching her forties and had no children, had just lost her job, and who finished the conversation by saying: "You know all this isn't so important, what really counts is my health." In order to establish to what extent this woman was justified in her assertion, it would be necessary to understand her personal conception of health. It is possible, however, to see here a general societal tendency to take a vague notion of health, inflate it in content and extension, and to make it the central point of life. In this way, the hospital has become heir to the church as the central institution of western culture (Rieff, 1979). We arrive at a societal deification of health. Health becomes an end in itself, the highest aim and the greatest good of life, i.e., it becomes a kind of quasi-religion.

c) It has already been noted that it is contradictory to understand health as a fundamental right of all persons. How does it stand with the opposite of this? Is health a duty? Simple as it is to come up with an answer in the case of whether health is a fundamental right, it is correspondingly difficult and complex to give an answer to whether health is a normative concept. If it is, then we are confronted with further questions, such as what kind of duty is health, is it a social duty, and can society demand health of its members? Is it a moral duty, is one obliged over and against oneself to be healthy? For how much of my own health (and illness) must and can I be responsible? Can something be a duty for which I cannot bear some or all of the responsibility? Here also, we find opening up a further field of questions, whose answers are central for a comprehensive understanding of health (cf. Parson, 1981; Szasz, 1978; Gorowitz, 1978).

6) To say that health is a state is at least incomplete, and shows a lack of understanding life in general and the human being in particular. It means forgetting that every living organism is generally not in a state, but is most of the time changing, evolving and functioning.

This must suffice for the criticism of the content of the WHO definition. The following criticisms do not concern the content but the manner and the spirit in which the definition was created.

B. Methodological criticisms of the WHO definition

I would like to point to some problems with the manner in which the task of drafting the definition was carried out, that is, with the methodological underpinnings of this definition.

First, to give a definition means to capture the essentials of a thing, person or living being, in a few words or signs. The problem is that the more complex the object to define, the more difficult it is to do so with a few words or signs.

It is still relatively easy to define something as simple as gravity, for instance. But what is the definition of a human being and what is the definition of a healthy human being? To my mind, there is no way to define a human being without trying to understand and describe him as completely as possible in all his dimensions. This means first studying and describing a "human being," followed by doing the same with a "healthy human being." After this there needs to be an analysis of the differences between the health of human beings and the health of other living beings. It is only at this moment that the essential findings of this process can be condensed into a formula or definition. The definition derived from this process, even though it might be short, stands on a big background. Depending on the specific situation, it might be necessary to go back to the basic description in order to clarify some aspects of health or solve a practical problem.[4] All this means a lot of intellectual work, and the WHO just did not do that work. It does not give us any background knowledge of how a human being must be viewed and understood. Neither does it give us references to where we can find such information.[5]

So, in summary, the WHO health definition is methodologically problematic because it seeks to define something that has not first been described. You cannot define something you do not even know.

A second important criticism can be introduced by the following sentences:

> I would like to beg you ..., to have patience with everything that is unsolved in your heart and to try to cherish the questions themselves, like closed rooms and like books written in a very strange tongue. Do not search now for the answers which cannot be given to you because you could not live them. It is a matter of living everything. Live the question now. Perhaps you will gradually, without noticing it, one distant day live right into the answer (Rilke, 1945, p. 21).

This implies also that questions of importance must be cherished, treated with patience or treated "philosophically," in the literal sense of the word. The WHO did not seem to have this necessary affection towards the problem of defining health. There was too much of a hurry. It could not even consult the scientists, philosophers and authors of antiquity. Otherwise, it would have seen that, for example, nearly two thousand years ago Galen wrote a complete theory of health, divided into a general (or theoretical) and a specific (or practical) part. This work is soundly based on the complete ancient understanding of human beings. Galen, for example, gives an explanation of why a word like "complete" is not useful for describing how healthy human beings function, a discussion on the problem of evaluating health in old age, and many things more. Also Hufenland in the 18th century wrote a complete description of health based on the vitalistic theory. This widely known book called *The Art of Prolonging Human Life* (1853) shows, among other things, that health is not a fundamental right, but an art. In our century, Sigmund Freud (1953, p. 269) wrote a useful definition of health, defining it as "a person's capacity for enjoyment and active life." This definition goes back to the Medieval monastic slogan of *ora et labora*, which again is founded on a complete Christian system of understanding the human being. But neither to these, nor any other historical source, do the documents of the preparatory commission make reference, giving the impression of a definition created *ex nihilo*.

IV. IMPLICATIONS FOR A NEW CONCEPT OF HEALTH

To my mind, the following can be learned from what was said above: First, the question of health calls for intensive, long-term scientific research done by scientists who are deeply motivated to solve this problem, but who are also patient enough to let the question be solved within our time and social context. As an example, let us take the problem of the growing shortness of financial resources, and as a consequence of this, the creation of new insurance models such as the so-called "Managed Care Models." By turning from paying for fees to paying for achieved results, there will be more and more interest in knowing how exactly these results can be defined. This is an argument in favor of encouraging managed care.

Second, in order to know more about health, we must carefully and respectfully study what others in the past have thought and written on this subject (van Spijk, 1991). Finally, we must formulate and systematize contemporary knowledge about human beings, and about healthy human beings. This also means that we need to create models of man, and that these models must be comprehensive and flexible enough to explain a wide range of practical situations.

In this context, the following clinical example might be illustrative. In 1956, a young man was ordained a priest, and a few months later stricken by the last huge epidemic of poliomyelitis in Switzerland. He became a tetraplegic, and also had to be ventilated artificially. He was from then on cared for in a hospital, had to be washed and fed, and could hardly talk because he had to take off the respirator and close the tracheotomy-tube whenever he wanted to speak. However, he could move his hands a little bit. When I first met him, he had already spent 21 years in the same hospital. Every Sunday, he prayed a private mass in his room. He had also engaged a secretary to help him maintain a huge correspondence, giving encouragement and counsel to many people he knew. Many letters he wrote by himself with a special mechanism which supported his arms so that he could use an electronic typewriter. He was highly appreciated by the medical staff and the medical director visited him on his weekly rounds, talking with him mainly about art, especially about music. In summer, he went out with a special motorized wheel-chair which he drove himself. Once he was asked whether he considered himself healthy or not. The answer was a clear: "Yes, I'm mostly healthy."

The models created to explain health should be able to tell us whether such a person is healthy, and if the answer is yes, to give the reason for it.

V. PROPOSAL FOR A MODEL ABLE TO SYSTEMATIZE OUR KNOWLEDGE ABOUT HEALTHY HUMAN BEINGS

A. The four-axis system

To my present knowledge, the General System Theory is a useful instrument to help us order our present knowledge about human beings (von Bertanlanffy, 1962). Using this approach, I propose to locate the human being within four different axes.

These axes are as follows:

Axis I: Three different *dimensions* of the object, namely:
  1. state,
  2. function, and
  3. evolution/goal (*telos*) (Le Moigne 1983).

Axis II: At least five different *system levels*, namely:
  1. basic element,
  2. cell,
  3. organ,
  4. organism, and
  5. environment (Sobel, 1979, pp. 87 ff).

Axis III: Three different *levels of being*, namely:
  1. physical,
  2. biological, and
  3. self-aware (Morin 1977, p. 86).

Axis IV: Two different *perspectives of describing*, namely:
  1. objective, and
  2. subjective.

This model is intended to clarify, coordinate and frame whatever is said about health, in order to reduce misunderstandings and promote discussions and a true dialogue. It will, for instance, help to solve the question raised earlier of whether there is such a thing as pernicious health. It seems that the authors who question the positive value of health think of health as a biological phenomenon (axis III.2), and express their concern about giving an exaggerated importance to this level of being. At the same time, they generally stress the point of giving the specific human dimension which I call here 'self-awareness' (axis III.3) its due importance.

In this model, some of the relevant sciences for our question can be seen as follows. Histology and anatomy become important for axis I (structure), axis II (cells, organs and organism), axis III (the biological level), and axis IV (the objective perspective). Human psychology becomes important for axis I (function, evolution and structure), axis II (organism and environment), axis III (self-awareness), and axis IV (objective and subjective perspectives).

Physiology, sociology, embryology, etc., can be classified in a similar manner. Parts of this framework are well known. Physiology is concerned with the proper or healthy working of the human body, anatomy is concerned with the proper or healthy structures within it, and statistics helps to understand where the limits of bodily health are, limits which often are not clear-cut, but show a more as less wide "gray zone" representing the neither normal nor diseased. Other parts of this framework are not so well known. By acknowledging this fact, we enter less explored parts of the above mentioned framework, namely the dimension of *telos* (axis I.3)[6], the significance of society (axis II.5)[7], and the significance of subjectivity (axis IV.2).[8] These topics must not only be treated and understood in themselves, but also in their interrelations with all the other topics of the framework. An important contribution to the understanding of the complex interactions within each one of the four axes is found in the work of Morin (1977, 1980, and 1986).

In the following, I would like to focus upon some points of this proposed four-axis system that to my mind seem important. The first is the relation the three elements of Axis III have to each other. After this, I would like to make note of the importance of the boundaries between the various elements within one of these axes. Special attention shall be given to the boundaries between "objective" and "subjective" in axis IV.

B. The three levels of being (axis III) and the significance of self-awareness

Our world seems to exist within three different existential levels: first and basic for everything are the systems of inanimate matter. They are necessary but not sufficient conditions for all living systems or living organisms. There is no life, no biological level of being, without a sufficient quantity and quality of inanimate matter.

Every form of life has its proper psychic dimension. That is, the psychic as I understand it is not a dimension added to life, but is always simultaneous with it. Even a simple organism such as a bacteria has its own 'psychism', which means that it 'knows' about its needs, fights its enemies, attacks its prey and so on. It is *"a knowledge which ignores itself"* (Morin, 1980, p. 184). This basic form of "knowledge" and "feelings" are phenomena common to all life, and belongs to what I call the biological level.

Kass gives an illustrative description of health in the biological sense, as I mean it here:

> What, for example, is a healthy squirrel? Not a picture of a squirrel, nor really or fully the sleeping squirrel, not even the aggregate of his normal blood pressure, serum calcium, total body zinc, normal digestion, fertility and the like. Rather, the healthy squirrel is a bushy-tailed fellow, who looks and acts like a squirrel; who leaps through the trees with great daring; who gathers, buries, covers but later uncovers and recovers his acorns; who perches out on a limb cracking his nuts, sniffing the air for smells of danger, alert, cautious, with his tail beating rhythmically; who chatters and plays and courts and mates, and rears his young in large improbable looking homes at the top of the trees; who fights with vigor and forages with cunning; who shows spiritedness, even anger, and more prudence than many human beings (1985, p. 173).

Thus far the problem of health seems to be solved for squirrels, but not yet for human beings. This is because there are fundamental differences between animals and human beings. On the basis of this, Kass has a definition of health that runs thus: "Health is a natural standard or norm ... a state of being that reveals itself in activity as a standard of bodily excellence or fitness, relative to each species and to some extent to individuals, recognizable if not definable, and to some extent attainable" (1985, p. 173).[9] Health as it is here described by Kass reflects axes I.1-3, II.4, III.1, 2 and IV.1 of the four-axis system presented above.

Insofar as Axis III is concerned, we find in Kass's description of the healthy squirrel all the elements of the biological level of being: important bodily (somatic) elements, and in addition to these, psychic characteristics (such as "courts and mates, fights with vigor and even anger"). These characteristics also belong, to my mind, to the biological level of being. The psychic is, however, in the case of the squirrel (and in the case of any kind of life other than human life), not aware of itself. Kass' description and definition correspond, in other words, to axis III.2. These characteristics, however, do not suffice to describe and define human life (and therefore human health), because in the case of human life, there is a third level of being. Perception, feeling, thinking and remembering, each of which are aware of themselves, are specific to human beings. I would like to call this self-awareness.[10] Here again there is no self-awareness without a sufficient quantity and quality of cells and organs, i.e., animate matter. So, we can view the phenomenon of self-awareness as the top of a pyramid. The top necessarily depends on its physical and biological basis. In our early life, the three levels of being are consecutively built one after the other.

But physics (inanimate matter) and biology (animate matter) are not sufficient elements for the constitution of self-awareness. Long, intensive interpersonal interaction is necessary to create it during childhood. It is something that evolves over the course of years, and which will take a central place in the existence of every human being. It is intrinsically connected with human language, culture and the knowledge of our inevitable individual death; self-awareness is also the starting point of our religious and spiritual life. A definition of health that leaves unaddressed the dimension of self-awareness, that important dimension of the human person, cannot to my mind fulfill the demands of theoretical coherence and practicability.

C. The boundaries between the subjective and objective perspective in Axis IV

The rational for drawing the limits between axis III.2 and III.3, i.e., between the biological and self-aware level, needed some supplementary explanations; this is also the case for the distinction between the objective and subjective perspective in axis IV. The apparent dichotomy between objective and subjective can easily be recognized as a gradual transition within a gradual shadowing. At the one end of the scale we find the subjective feeling of being healthy or ill which is purely based on one's subjective perception and not subject to any external control. This would be the case, for instance, for a patient who suffers from severe pain without any other objectionable bodily alterations. But this example shows also that this subjectivity is not yet a pure one: all of us know from our own experience what it means to suffer pain. There is a possibility of intersubjective communicability on the ground of common similar perceptions and experiences in the past. Pure subjectivity, however, is not communicable because the act of speaking already transforms a purely subjective experience into a more objective one. (It goes without saying that there might be pure subjectivity and pure objectivity, but that they are not of any importance for our more practical question of the nature of health and disease. There is probably a pure subjectivity but it cannot be communicated, and there might be a pure objective stand toward our world, but we cannot know it).

The transitional steps from a pure subjectivity to a pure objectivity may be seen as follows:

a) subjective experience that is pure and incommunicable;
b) subjective experience that is difficult to communicate and can be understood by very few other human beings only (e.g., an experience of religious enlightenment);
c) subjective experience that can quite easily be communicated to a more or less extended group of other human beings (e.g., pain or pleasure). An experience becomes more objective the bigger the group of people becomes to whom it can be told and who can understand it;
d) the subjective experience goes together with observable and measurable changes in function and/or structure of the organism. These changes can be subtle (e.g.; a slight stress-reaction of a person recognized by close family members only) , or it can be included in the world-wide spread of literature of a well-known disease (there is not yet much literature describing healthy states);
e) objective changes can be observed but the person feels subjectively unchanged (e.g., carcinoma *in situ* of the cervix). Here again the degree of objectivity increases the bigger the number of people who share the same observation;
f) an objective authority who stands outside of the human community is giving the standards for what has to be seen as healthy or sick (e.g., nature or a natural norm). Human beings are, however, able to have an understanding of the plan of this authority;
g) pure objective knowledge of the essence of human health by an exterior authority out of the range of human understanding.

The limit between objective or subjective, in my mind, must be drawn between c) and d) in the list above. Among the authors who give more importance to subjective criteria we can mention C. Withbeck, who describes health as follows: "People generally recognize the value of having the psychophysiological capacity to act or respond appropriately in a wide variety of situations. By 'appropriately,' I mean in a way that is supportive of, or at least minimally destructive to, the agents goals, projects, aspirations, and so forth" (1981, p. 611).

L. Nordenfelt introduces a supplementary objective criteria introducing the term "vital" goals in his definition of health. He says: "A person's health is characterized as his ability to achieve vital goals" (1991, p. XI). H.T. Engelhardt (in this volume) takes an objective stand that also has subjective components. This is because he assumes that every social group creates its own (objective) understanding of health and disease which again is largely dependent on the subjective position toward the world and the human beings taken by this group. L. Kass, in his definition, refers to an objective exterior (natural) norm. He says: "health is a natural standard or norm ... a state of being that reveals itself in activity as a standard of bodily excellence or fitness, relative to each species and to some extent to individuals, recognizable if not definable, and to some extend attainable" (1981, p. 18).

D. Delimitation of disease within an isolated part of the system

Applying General Systems Theory includes also the use of its principles of steady state that helps the system to resist destabilizing influences. Even though these principles are not well investigated and conceptualized on a medical level (Le Moigne, 1983, pp. 197 ff.; Sheldon, 1970, pp. 84 ff.), it can basically be said that as long as a problem can be restricted to any part of the system (for instance within an organ (Axis II.3), or within the structural level (Axis I.1) without affecting the function, and so on), health is still possible, even though it becomes more fragile.

VI. APPLICATION OF THIS THEORY TO OUR CLINICAL EXAMPLES

My hypothesis is the following: The paralyzed priest mentioned in section IV had already become a highly self-aware person when he became paralyzed. At the moment of paralysis, parts of the physical and biological basis of this self-awareness broke down, but the top of the pyramid, the core of existence, namely self-awareness, stayed intact. The person was able to *restrict the problem* to a biological level and was able to *pursue his existential goals*. To my understanding this person is still healthy.

The second example concerning the lady who believed that health was the most important thing in life raises the question of the proper goals of human beings. This question certainly needs further investigation. The answers given will tell us whether this person can be considered healthy or not. Without entering into a more detailed discussion now, I'd like to agree with V. Frankl (1972, p. 75, author's translation) that "The essence of human existence lies in its self-transcendence."[11] Therefore, I think this woman cannot be called healthy, and she runs a high risk of becoming diseased in the near future.

With regard to the example of the homosexual woman given at the beginning of this article, the following points seem important: socially, a deviation from the given norms, such as homosexuality, can produce so much pressure that the feeling of subjective well-being becomes practically impossible. In the present example this seems not to have been the case, because subjectively, the person felt reasonably well. Somatic health, in the sense in which western medicine understands it, is perfectly compatible with homosexuality. Up to this point, I have not been able to find any reason why homosexuality *per se* could substantially affect self-awareness. I still believe that stopping treatment after the grief reaction had passed was the right thing to do.

Lucerne, Switzerland

NOTES

[1]   Annex 9, Appendix. I have chosen to work with the French translation of the documents, since the English translation differs from the French on what are to my mind significant points, and the English text given in the paper are my own. The official English document renders this text as follows: "Health is not only the absence of infirmity and disease, but also a state of physical and mental well-being and fitness resulting from positive factors, such as adequate feeding, housing and training" (WHO, 1946b).

[2]   Annex 10. The official English document renders this text as follows: "Health is not only the absence of infirmity or disease but also a state of physical fitness and mental and social well-being" (WHO, 1946d).

[3]   Cf. WHO: "Health promotion is the process of enabling people to increase control over, and to improve, their health. To reach a state of complete physical, mental and social well-being, an individual or group must be able to identify and to realize aspirations, to satisfy needs, and to change or cope with the environment. Health is, therefore, seen as a resource for everyday life, not the objective of living. Health is a positive concept emphasizing social and personal resources, as well as physical capacities. Therefore, health promotion is not just the responsibility of the health sector, but goes beyond healthy life-style to well-being" (1986).

[4] To take the example of gravity, the definition

$$G = k(m \times M/r^2); \; k = 6.67 \times 10^{-11} \; m^3kg^{-1}s^{-2}$$

is useful most of the time, but in special situations, such as the gravity of atomic structures, we need to reconsider the exact meaning of k or m.

[5] In all the records of the committee meetings of the WHO available to me, there are no references to deeper discussions of the epistemological grounds for this definition, and there are no references given to sources upon which the proposed definition is based. (See e.g., WHO, 1946 a,b,c.)

[6] What is the proper aim and purpose of a human being? See M. Lavados' contribution in this volume.

[7] What is the society's influence on the establishment of goals and proper functioning in an individual life? Has health to be defined differently in every society? See H. T. Engelhardt's contribution to this volume.

[8] See P. Donohue-White's and K. Fedoryka Cuddeback's contribution in this volume.

[9] This definition and the corresponding argumentation is very similar to that of Galen in his *de sanitate tuenda*.

[10] With the suggested subdivision of Axis III into physical, biological and self-aware, various concepts take on a different connotation: the bodily and somatic (Axis III.2) are distinguished from the physical (Axis III.1). In the first there is living matter, in the second non-living. The psychic belongs, as already noted, to Axis III.2 also. The distinction between somatic and psychic has a long tradition, but is nevertheless often arbitrary, and in practice its usefulness is questionable. The word "psychic" is certainly often used differently in practical usage than I use it here. For instance, psychiatry and psychology in particular are concerned primarily with problems whose origin is to be found in a disturbed self-awareness. As a consequence of disturbed self-awareness, not only pathologically altered psychic phenomena (e.g., ungrounded anxiety as the result of an *unconscious* childhood experience), but also somatic disturbances, can proceed from the same cause. Self-awareness as it is here understood has, among other connections, a strong relation to the "I" of the psychoanalyst.

[11] Reliable opinion polls invariably tell us that a majority will agree in calling health the most important thing in life. I believe (and this belief is consistent with the Platonic understanding of the world) that there are basic realities standing above the spheres of society and cultural groups.

BIBLIOGRAPHY

Bertalanffy L.v.: 1962, *General System Theory*, George Braziller, New York.

Callahan, D.: 1973, 'The WHO definition of "health",' *Hastings Center Studies* 1(3), 73-88.

Engelhardt H. T.: 1981, 'Concepts of health and disease,' in A. Caplan and H.T. Engelhardt (eds.), *Concepts of Health and Disease: An Interdisciplinary Perspetive*, Addison-Wesley Publishing Company, London.

Frankl V. E.: 1972, *Der Mensch auf der Suche nach Sinn* (original title: *Das Menschenbild der Seelenheilkunde*), Herderbücherei, Freiburg.

Freud S.: (1953 [1904]), *Freud's Psychoanalytic Method*, The Hogart Press, London.

Galen C.: 1951, 'De sanitate tuenda,' in R. M. Green (trans.), *A Translation of Galen's Hygiene (De Sanitate Tuenda)*, Charles C. Thomas, Springfield, Illinois.

Gorowitz, S.: 1978, 'Health as an obligation,' in W. T. Reich (ed.), *Encyclopedia of Bioethics*, The Free Press, New York, pp. 606-609.

Hersch J.: 1981, 'Santé: l'utopie d'une definition,' in *Schweizerische Aerztezeitung* 62, 1531-1532.

Hufeland C. W.: (1853 [1796]), 'Hufeland's art of prolonging life,' E. Wilson, London.

Kass L.: 1985, *Towards a More Natural Science*, Free Press, New York.

Le Moigne, J.L.: 1983, *La théorie du systéme général*, PUF, Paris.

Morin E.: 1977-86, *La Méthode* (1977, *La nature de la nature*; 1980, *La vie de la vie*; 1986, *La conaissance de la conaissance*), Le Seuil, Paris.

Parson, T.: 1981, 'Definitions of health and illness in the light of American values and social structures,' in A. Caplan and H.T. Engelhardt (eds.), *Concepts of Health and Disease: An Interdisciplinary Perspsective*, Addison-Wesley, London, pp. 57-82.

Rieff, P.: 1979, *Freud - The Mind of the Moralist*, 3rd edition, Chicago University Press, Chicago.

Rilke R. M.: 1945, *Letters to a Young Poet*, Sidgwick and Jackson, London.

Schadewaldt H.: 1965, *Die Apologie der Heilkunst bei den Kirchenvätern*, *Veröffentlichungen der internationalen Gesellschaft für Geschichte der Pharmazie*, Stuttgart, pp. 115-130.

Sheldon, A. et al.: 1970, *System and Medical Care*, MIT Press, Cambridge, Massachusetts.

Sobel, D. and Brody, M.: 1979, *Ways of Health*, Harcourt Brace Jovanovic, New York.

Spijk, P.v.: 1991, *Definition und Beschreibung der Gesundheit*, *Schriftenreihe der SGGP N. 22*, Muri, BE, Switzerland.

Szasz, T. S.: 1978, 'The concept of mental illness: Explanation or justification,' in H.T. Engelhardt and S.F. Spicker, eds., *Mental Health: Philosophical Perspectives*, D. Reidel Publishing Company, Dordrecht, pp. 235-250.

WHO: 1946a, 'Actes Officiels de l'Organisation mondiale de la santé,' no. 1.

—— 1946b, 'Official Records of the World Health Organization,' no. 1.

—— 1946c, 'Actes Officiels de l'Organisation mondiale de la santé,' no. 2.

—— 1946d, 'Official Records of the World Health Organization,' no. 2.

ROCCO BUTTIGLIONE AND MANUELA PASQUINI

THE CHALLENGE OF GOVERNMENT IN THE CONSTRUCTING OF HEALTH CARE POLICY

I. INTRODUCTION

In the political arena, we are used to considering and hearing about health issues from a monetary perspective. To a certain extent, this perspective is justified and understandable, for our democratic states – especially the most developed countries of the West – are confronted today with the real danger of financial collapse. Indeed, one of the main reasons for the financial crisis of our democracies is the constantly growing expense of health care. It seems that the cost of health care is expanding and will expand at a pace that will make a balanced governmental budget impossible in the near future. Consequently, one of the most difficult challenges for any government is to find a way to reduce health care costs without violating what is more and more frequently referred to as the right to health.

The main purpose of our contribution to this volume is to point to three philosophical problems that underlie the political and economic problem of developing and financing state health care, namely,

a) The question about what health is;

b) The question of whether it is a right; and

c) The question of what the content of such a right would be.

Possible answers to these philosophical problems will be presented, and to some extent evaluated, from the viewpoint of the taxpayer. We invite you to a kind of paradigm shift. Until now, questions of bioethics and the philosophy of medicine have been approached primarily from the point of view of either the physician or the patient. Our challenge is to lead you to consider the perspective of the tax-payer – i.e., of the one who, in the final analysis, finances that which is called for when we speak of a 'right to health'.

P. Taboada, K. Fedoryka Cuddeback and P. Donohue-White (eds.), Person, Society and Value: Towards a Personalist Concept of Health, 229–239.
© 2002 *Kluwer Academic Publishers. Printed in Great Britain.*

II. HEALTH CARE: ONE OF THE HIGHEST EXPENSES IN GOVERNMENTAL BUDGETS

As we already mentioned, the costs of health care have been skyrocketing and absorbing more and more governmental resources. The phenomenon of the growing expenditures in this field is due primarily to both the development of medical knowledge and a potentially unlimited demand for health assistance. Today we can treat diseases we were unable to treat not only twenty, but even five years ago. The development that allows us to cure illnesses that once could not be treated calls for programs of medical research. Thanks to this research, we both discover new medications and develop equipment that can treat a given illness, or perform a certain operation. In the experimental stages of this development, the access to new therapy cannot be possible for all; in fact, many must be excluded. Such a circumstance poses a remarkable dilemma, namely, the choice – and the criteria of such a choice – of which persons are entitled to this treatment. (Presently, there is a passionate debate on this topic in the United States.)

Further, whatever the solution might be at this experimental stage, equal distribution of this treatment makes it necessary to spread the new knowledge, to provide hospitals with the new instruments, medications, and so forth. Of course, the whole process of training medical practitioners and providing hospitals with the appropriate means for more effective therapy costs a lot of money. These empirical considerations are linked to the practical question of who is to pay for this development, or, in more general terms, how we shall finance our health care expenditure.

Thus, we see the problem – technology has grown so much that the equal distribution of health care is almost impossible to finance. This is a financial, and ultimately a "tax-payer" problem. Let us now point out the various philosophical problems involved in this seemingly "economic" issue. The issue involves, first, the attempt to understand and define what health actually is. Then, we can proceed to investigate whether a right to health in fact exists, determine the content of this right, and finally, figure out who should pay for it.

III. HEALTH, EASIER SAID THAN UNDERSTOOD

The politician, in addressing this issue, could stop to ask himself this question, "What exactly is this thing I want to finance?" He could take the

World Health Organization definition (1958) for his answer: "Health is a state of complete physical, mental and social well-being and not merely the absence of disease or infirmity."

This definition does, in fact, regulate some of the relations between states. For instance, a significant part of the expenditure for health in the world is today channeled through inter-governmental agencies that allocate the resources of the richer countries to the poorer countries, in order to provide them with health care. The WHO definition was drafted precisely in response to the need for such international efforts for a definition of what it was that these efforts sought to achieve.

It is needless to say that a clear understanding of the nature of health is indispensable in determining what can and what cannot, what should and what should not be comprehended in the action of a health care organization. For example, should abortion be financed as an element of health care or not? And if it should, then should childbearing be considered an illness, or an accident? We leave the answers to these questions open to your reflection, and of course, there are other examples that could be offered. But any answer to this or similar problems will certainly begin with the question of whether and how the problematic issue is considered as part of or related to health.

Becoming aware of the existence of the WHO definition does not, however, give the politician a quick answer to this first question. The politician is faced with the further problem that the WHO definition allows for various interpretations. Let us picture a case which, by virtue of its simplicity, can better clarify both the limitations of the WHO definition and the paramount importance of a clear understanding of health and what health care should comprise. Let us imagine that we find a way to increase the growth of young men. In a society in which this is possible, anybody could say, "I wish to be six feet tall, and I have a right to receive this medication that allows me to grow to six feet." Would such a claim be legitimate? And then one could also ask, why six feet and not six feet three or five feet ten? What is the criterion for where the line should be drawn?

The WHO definition of health does not help much in finding an answer to such a problem. It tells us that health is "complete physical, mental and social well-being." The meaning of this rests and depends on what we understand by 'well-being'. And there are several possible interpretations of this: emotionalistic, materialistic, and realistic.

An *emotionalistic* understanding of the definition considers health as the exclusion of suffering; it identifies what feels good with what actually is good. Now, it is evident that the relief from pain is one of the goals of medicine and that in many instances feeling well also means being healthy, but this is not always the case. There are indeed exceptions, which point to the fact that the way one feels cannot be the ultimate criterion of one's welfare.

From a pragmatic standpoint, if well-being were the same as feeling well, then one would have a right to have whatever was supposed necessary in order to feel well, for instance, being six feet tall. But not all that feels good *is* good. One of the most dramatic instances in which we become aware of this fact is that of drug addiction. Drugs make you feel good, at least for a while, but they are not good – they can kill you. This should lead us to consider that the subjective measurement of health through feeling well is at least doubtful. Not only in ethics, but also in medicine and with regard to health, there is a distinction to be made between what one subjectively feels, and what really is the case. What I subjectively feel as being good is "relative to me"; it corresponds to my likings and desires. It is not the result of an acknowledgment of the nature of the thing itself, but rather, an estimation of how it "fits" my present wishes. Although my personal judgment and my subjective emotions are, as a rule, a good index that leads me to reality, it is not always the case that they in fact do so. For sometimes, as we mentioned, I can feel that something is good, while in fact it is not. Moreover, if we accept an interpretation of well-being that is subjectivistic in this sense, it could be very dangerous: dangerous for the patient, and equally dangerous for the tax-payer, who, in this specific case, should pay for the drugs.

Admittedly this is not a philosophical example. It is a matter we are discussing in the Italian Parliament, since one proposal of how to treat drug addicts is to give them drugs, so that they will be happy, they will not steal, they will not weep, they will not cry and we will not fear them. Of course, they will die after a while. But this is the greatest possible reduction of damage – and reduction of damage in this area sounds good. Thus runs the well-known and well-accepted argument of the lesser evil! But, on the other hand, if we have a non-relativistic interpretation of well-being, then we must respond with a "no." Our response is not to provide the addicts with drugs, but rather with a chance to get them out of their dependency on drugs.

A *materialistic* definition of health has also revealed itself to be inadequate. It considers the body as a machine, according to the old tradition of Descartes and La Mettrie, with its own, mechanistic rules of functioning. Accordingly, health consists in the physiological functioning of the body, as if the human being were only body. It ignores completely the psychic part of the person, ergo it cannot take into account psychological illness. Considering the human being as an object, it forgets that it is also a subject, and that this being a subject is more objective than anything. We experience this subjectivity in conscious action, in deciding to act or not to act, in our self-determination and self-possession, in reflecting on our lives, in the "voice" of conscience. Being a subject, with all that this entails, is such an undeniable characteristic of the person that it is actually constitutive of it.

The materialistic conception was widely accepted until, we believe, the thirties. The ideology of medicine changed after the first World War, when it was acknowledged that psychological diseases are as cumbersome as purely physical diseases. Indeed, if we measure social illnesses with the amount of working time that is lost because of them, then we see that psychological illnesses take a large share of the working hours that are lost in one year in countries like Italy, the United States, or the like. Indeed, a headache, for instance, is an illness, and we are unhealthy because of a headache, even if its cause is a psychological one. And, while the demand to be six feet tall in order to be happy is unreasonable, the demand not to have a headache is reasonable.

There is an aspect in medicine that transcends what is reflected in the impoverished materialistic outlook, but this aspect cannot be extended subjectively, in the sense that whatever I wish and consider as pertinent to my well-being really *is* pertinent to my well-being. The introduction of the subjective element can indeed lead to the opposite danger of transforming this position into a subjectivistic one, which brings us to the *realistic* definition of health.

What we want to call the "realistic interpretation" of health combines, so to speak, elements of both the preceding interpretations. In this understanding, health is a property, a quality of an object that functions in the same way in which we say that a car works and that a machine functions. It is a mechanical functionality. But health pertains to an object, which happens to be a subject, and which has a subjective side that is linked to the object in a way that is not arbitrary.

The problem of understanding health comes very close to the general problem of modern philosophy, that is, to introduce subjectivity without becoming subjectivistic. In other words, it is the problem of how to consider subjectivity as an objective component of the human being. We believe this is the path we should take in order to find a definition of health that is philosophically sound, and that is at the same time "operational," as they sometimes say (in a very bad neologism).

Let us return quickly to our previous examples: abortion, euthanasia, and drug consumption. If we tackle these issues from a realistic perspective, our "no" to them becomes clear, since how we feel about them (I feel that a child is a burden, therefore abortion; that my life is meaningless, therefore euthanasia; that I need to forget my problems for a time, therefore drugs) is not the ultimate criterion. Nor is the functioning of the body the ultimate criterion (the child-to-be will be handicapped, therefore abortion; I am not self-sufficient any longer, I do not "function," therefore euthanasia).

Going back to the thing itself, the realistic approach takes into account the whole: how our body works, how we feel and the particular way in which these two are related to human health, which is more than just a combination of the two. The realistic definition of health is both adequate to reality, or our common experience of reality, and the most suitable for health care policy. In fact, common sense does not find life something to be annihilated when a person cannot function "properly," or cannot enjoy life for psychological reasons. A realistic definition further assures that neither dimensions of health – the subjective or the objective – is emphasized at the cost of the other. It recognizes that psychological and spiritual aspects influence a person's ability to function, and also directly affect her physical health.[1] At the same time, it can avoid the arbitrariness of an emotionalist vision, which alone gives rise to endless demand, because of the limitless expectations of the subjectively felt need.

What we have here is an approach that delineates health firmly enough so that the tax-payer and the legislator know up front what they are paying for, and can have at their disposal certain fixed criteria for evaluating the way their money is spent. Without such a balanced conception, the state – if it is the one to provide health services – risks bankruptcy. The realistic approach can be the basis for sound political and administrative action, because it does put a limit on possible claims. We would even say that it could be the basis for sound administrative action, because it is philosophically sound.

The "good health" of our state budgets, which is for the politician a fundamental source of worry, depends on how the health issue is tackled. We think that a correct political approach takes into account both the nature of health and the right to health. Presupposing a realist account of the nature of health (which we have not provided, but only point to as the best approach), the legislator must turn to the question of whether and how it is a right.

IV. THE RIGHT TO HEALTH

The politician, in tackling health care policy, might well ask himself if this is in fact part of his task, if health is something to which everyone has a right, and if so, then whether it is *his* duty to ensure its supply. This brings us first to the philosophical question of what a right is at all, which is a huge question. We will here only point out some of the elements of this question relevant to the question of health. The main points to be addressed are two: a) the relation of rights to duties, and b) the identification of the bearers of these duties.

Von Hayek (1978, 1978a, 1981) poses the question of how one can say that anybody has a right to health, when one cannot determine the corresponding duties. According to this line of thought, if one has a right to health, then, there must be somebody else who has a duty to preserve this health. But if we do not, or cannot, determine who is to carry the burden of implementing the right, then we cannot speak of rights.

In the specific case under analysis, we might be puzzled by the question of who has the duty to keep us healthy. Possible answers are: God, nature, or the State. However, God and nature cannot be subject to our laws. God is sovereign, and so we cannot impose a duty on Him. Neither can we impose it on nature. The State, however, does not have the power to implement this right; it cannot give us our health – it can at best help us to a certain extent if we need help in regaining it. Faced with this inevitable state of affairs, we might come to the conclusion that there is no right to health, or no right in this sense.

Here, we think, the problem regards not only health, it regards all social rights. We should perhaps define this kind of right more concretely as a right to the cooperation of other individuals in the implementation of the state of affairs that is the object, or the content, of the right. With this approach, we can, then, find a sense in which we can speak of a right to health, or rather to health care, a sense that overcomes the criticism of von Hayek. Indeed, while we cannot say that anybody has a duty to keep us healthy, we can say that we have a right not to see our health endangered through the action of others. This would be accepted by von Hayek, who holds that general rights are always negative rights. We may have a general right to be allowed to perform a certain action, and this would imply a general duty not to interfere with this action.

But we can say something more. We can speak, as we already mentioned, of a right to the cooperation of others in the implementation of social rights.

According to this conception, the primary carrier of the responsibility for the implementation of the right is myself. We cannot shift totally to the shoulders of others the task of having our rights implemented, while we *can* expect others not to interfere with that task. Here we must consider the question of how to identify these others.

A certain trend among modern liberals is to take the Austrian-American School of Economics as point of reference. They would agree with what Ludwig von Mises (1949) would presumably say, i.e., 'you are the first, foremost, and only carrier of this responsibility; it is your own business'. We know, however, that in reality such a stance does not work, because there are circumstances under which the burden can be too heavy for the individual, who, unable to withstand it, would be left to drown in a situation of despair. This would be, on one hand, inhuman, and, on the other, socially and politically dangerous.

Another answer is that of Lord Beveridge, according to whom the State is the sole carrier of the responsibility. Following his view, we should build up a national health system that would take care of every individual's medical needs. We know that this also does not work. It costs too much and, in addition, one never clearly knows what should be given and what should not be given. In the long run, the outcome is very poor assistance, at very high costs, all seasoned with a growing dissatisfaction of the people.

The political problem that arises when government takes on the responsibility of implementing the individual right to health is the determination of what should be given and what should not. If the right to health were a merely negative right, it would entail only that nobody should injure us. No positive claim could be deduced from it. But social rights, as we have already suggested, have a positive element too. When we commonly speak of the right to health, we usually mean and refer to a service we think we are entitled to: health assistance. And what health assistance comprises is the big question coming back again and again.

As we have already proposed, a correct political approach should be a realistic one, which takes into account the complex nature of both health and the right to health. Such an approach should put a part of the responsibility these entail upon the individual, and a part of the responsibility upon civil society. By this we mean not only the network of non-profit organizations and associations[2], but also for-profit insurance companies, which can either take over a part of the burden, or be an instrument for taking a part of it. Laws can play a very important role in helping people to assume their share of the responsibility, by obliging them to pay for the social security of their family or for health insurance for the needs of their family. Then there is, of course, the responsibility of the State. A realistic discussion of health care politics must apportion how much of the burden should be carried by the individual and the State. Any attempt to reduce the load to only one subject, on one hand, does not correspond to the reality of today's world, and on the other hand, is conceptually biased. It does not correspond to the essence of social responsibility and to the essence of health.

Again, if we agree that not all the weight should be put on the State, then the problem is to define the role of the State. It seems that there is a basic sense of health, some fundamental elements of human health, which should be preserved under all circumstances, and the State must defend what we might call the "basics" of health. Such a general assertion might appear self-evident; however, a closer look at it reveals how hard it might be to define the limits of these basics. It seems obvious that basic treatments for physical diseases should be provided to any and each person, and also to those who cannot pay for it. The reason for this is the affirmation of the "other" as a human person, whose life and possibility of actualization of his full being ought to be preserved. But where does one draw the line, determining where this 'basic' responsibility ends?

We have not really been making an argument in favor of *a right to health* here. And we criticized this formula because nobody has a right not to feel sick, not to suffer an illness, not to die. We propose, instead, to substitute for the expression "right to health" the expression "right to health care." If one is sick, one has a right not to be left alone, to be attended to, and to receive all the support that can be given.

The problem we focus upon is: who has to pay for the implementation of this right? It seems that the responsibility for the implementation of the right to health care should be shared by the individual, the primary social group to which the individual belongs and, last but not least, the State. One should have medical insurance, if one can pay for it. In case of need, family, local communities, and non-profit organizations should take a share of the burden. If no other option is available, the State should intervene. Mixed systems in which all these levels of responsibility are actualized seem to correspond better to the nature of the right to medical care than systems in which the State is the sole provider of medical care.

Now we propose, but do not answer, another question: what are the limits of the right to health care? The technical means of providing a treatment go far beyond the organizational possibilities of putting all these means at the disposal of every human being in need. We must have criteria to select those who will have and those who will not have access to limited medical resources. The market criterion cannot be enough (those will be treated who have money to pay for the treatment). Other criteria should be developed (for instance women and children first). This is perhaps the most delicate and difficult problem not only of bioethics, but also of health care politics today. We know that all men have a *right to ordinary means* to restore health. We would not defend a *right to extraordinary means* for all. But it is of course very difficult to distinguish between ordinary and extraordinary means, and it is equally difficult to establish who has a right to extraordinary means and who does not.

V. CONCLUSION

We began our considerations with the puzzling question of whether a rationing of medical goods can be morally accepted (because, let us face it, the rationing of medical goods is a matter of fact). We offered many examples to illustrate the complexity of our subject matter. There are no easy, pre-packaged answers. Quite the opposite is the case, and this paper neither could unveil nor aimed at unveiling such complexity. What it has attempted to do is to show that this is a multi-dimensional complexity, combining political, philosophical and practical questions. And we hope it has become evident that in many cases, the political and practical solutions depend upon the elaboration of an adequate philosophical foundation.

To summarize our contribution, we could refer to a small book, very famous and very good, written by Rudolph von Jhering (1972): *Der Zweck im Recht* (*The Finality in Law*). The point is that in order to make good laws, we must make clear what is the finality, the goal, of the laws we make. And, if the goal is health, we need a realistic account of what health is.

International Academy of Philosophy
Principality of Liechtenstein

NOTES

[1]   See for instance the studies of Luban-Plozza (1978, 1992).

[2]   At times such associations are far better than State structures; just two examples: Alcoholics Anonymous and, in Italy, San Patrignano for drug addicts.

REFERENCES

Hayek, F.A.v.: 1978, *Law, Legislation and Liberty*, vol.1, *Rules and Order*, University of Chicago Press, Chicago.

—— 1978a, *Law, Legislation and Liberty*, vol.2, *The Mirage of Social Justice*, University of Chicago Press, Chicago.

—— 1981, *Law, Legislation and Liberty*, vol. 3, *The Political Order of a Free People*, University of Chicago Press, Chicago.

Jhering, R.v.: 1972, *Lo scopo nel diritto*, G. Einaudi, Turin, Italy.

Luban-Plozza, B., Egle, U., and Schüffel, W.: 1978, *Balint-Methode in der medizinischen Ausbildung*, Gustav Fischer Verlag, Stuttgart, New York.

Luban-Plozza, B., and Pöldinger, W.: 1992, *Psychosomatic Disorders in General Practice*, 3rd edition, Springer-Verlag, New York.

Mises, L.v.: 1949, *Human Action: A Treatise on Economics*, Yale University Press, New Haven.

WHO: 1958, 'Constitution,' in *The First Ten Years of the WHO*, Geneva.

JOSEF SEIFERT AND PAULINA TABOADA

EPILOGUE

At the end of this volume we would like to recall the point of departure
for the research project on the philosophical aspects of the concept of
human health that led to this book. Our purpose was to provide a critical
analysis of the WHO definition of health as "a state of complete physical,
mental and social well-being and not merely the absence of disease or
infirmity" (WHO, 1946), and to propose the outlines for a personalist
conception of human health. What are the results of our research? Which
are the main philosophical insights expressed in the contributions
gathered in this volume?

I. EPISTEMOLOGICAL, ONTOLOGICAL, AXIOLOGICAL AND CONCEPTUAL ISSUES

A first point of consensus that certainly emerges from the papers
collected in this volume is the fact that there is no concept of human
health that would not have, besides its obvious empirical aspects, definite
philosophical dimensions. And the first philosophical questions that need
to be addressed refer to the epistemological status, content, and
ontological referent of this concept.

A. Is Health a Positive or a Negative Concept?

The contributions to this volume all confirm the validity of an insight of
the WHO definition, namely the primary and *positive* character of health
and the consequent impossibility of defining it merely as the absence of
disease and infirmity. It seems quite clear that health is an eminently
positive good *of* and *for* the human person and that this positive character
precedes, at least logically and anthropologically, that of disease and
infirmity. These phenomena receive their negative character precisely in
contradistinction to the positive good of health; they diminish or destroy
health.

Though health often becomes the object of our consciousness and
gratitude often only when we lose it or when it is threatened through

P. Taboada, K. Fedoryka Cuddeback and P. Donohue-White (eds.), Person, Society and Value:
Towards a Personalist Concept of Health, 241–252.
© 2002 Kluwer Academic Publishers. Printed in Great Britain.

disease or infirmity, it is still evident that the 'hidden good' of health, as Gadamer (1987) puts it in his significant discussion of health, is in no way a mere absence of infirmity. Indeed, sickness and disease impose themselves far more conspicuously on our experience than the positive good of health. Nevertheless, we do also have a positive experience of what it means to be healthy. This point is brought out in various of the present contributions, particularly in Ide's, Lavados' and van Spijk's.

The positive content of the concept of human health also follows logically from the general metaphysical insight into the primacy of good versus evil. In fact, evil always implies some antithesis, destruction, lack, rejection, etc. of a preceding good. Thus, however difficult it is to state and formulate in language the content of health positively, health nevertheless is a positive phenomenon and must be regarded as such. Moreover, it presupposes the underlying positive phenomenon of life: it is the special flourishing and actualization of the life and potentialities of a living being, as Ide and Taboada emphasize. This has interesting consequences for the whole area of health-care that should serve not only the prevention of disease but, in the spirit of old and new dietetics, also the positive furtherance of health, as Reale suggests.

B. Is Health a Descriptive or an Axiological Concept?

But even if one were to admit a positive content of health as opposed to unhealth, one might still believe that the concept of health is merely descriptive and not value-laden. The idea contained in the WHO definition of health of a "complete well-being," is obviously the idea of a good or value, of a positive importance; and the ideas of disease and infirmity obviously imply some form of evil.

With Nordenfelt (1987, 1993, 1993a) and Engelhardt (in this volume), we recognize not a merely neutral descriptive but, rather, an axiological content of the notion of health. Moreover, this axiological content is a condition of any foundation of health-norms. Donohue-White and Cuddeback address this question, and the related problem of whether it is the very concept of health that indicates a value, or whether it is actually the reality designated by this concept what possesses value. They conclude that the concept of health has a content that is not reducible to a mere neutral description of the well-functioning of a being, but means at the same time a certain kind of good *of* and *for* this being.

All the authors in this volume seem to concur on a position according to which the notion of health is both descriptive and evaluative. On this issue we find not only a complete agreement among the authors of this volume, but also their agreement with the WHO definition, which clearly invokes a value-laden concept of health. Yet the WHO definition does not explain what exactly it means by 'well-being'. Therefore, the definition is open to any interpretation and would be compatible with both ontologically and epistemologically subjectivist notions of value. It is this vagueness of the WHO concept of health from the axiological point of view that becomes a major object of critical reflections in this volume.

C. Is Health a Subjective or an Objective Value?

Indeed, well-being could be understood as referring to a mere subjective experience of 'feeling well' of which we could neither say that it is *intrinsically good* nor that it is *objectively good for* the healthy person. The 'well-being' of being healthy could then be compared to the attractiveness of drugs for the drug-addict; it could be identified with the 'good' of something that is merely subjectively satisfying. The WHO definition of health does not exclude such an interpretation. If this were the case, well-being itself would fail to possess *ontological objectivity.*

But well-being could also be interpreted, along the lines of Mackie (1977), as presenting itself to us as objective but not actually fulfilling this claim to objectivity. A 'value objectivity' does not immediately follow from a 'descriptive objectivity'. For one could agree on the descriptive content of a conception of health but still deny the objectivity of its value. In this case, the ontological subjectivity of health would not allow us to say that its opposite, namely unhealth, is objectively destructive for a person, nor to distinguish the 'value' of health from the subjective pleasure that can be reached through drugs and alcohol. In fact, these substances could no longer be declared *objectively* bad or damaging for a person.

Two of the contributions contained in this volume explicitly address this aspect of the philosophical debate on the concept of health: the piece jointly written by Donohue-White and Cuddeback, who defend the objectivist view, and the piece by Engelhardt, who defends an epistemologically relativistic concept of health, while preserving the right of groups within society to adopt an *ontological* objectivity (for private or religious reasons).

All the authors of this volume agree on, and some of them seek to provide objective evidence for, the fact that the negation of the ontological objectivity of the value of health fails entirely to do justice to the phenomenon of health. An argument in favor of the objectivity of the value of health, which appears to be dominant in the paper on the politics of health by Buttiglione and Pasquini, is that a subjectivist notion of value would have destructive practical consequences for the organization of society, which no state is prepared to accept. Not only would there be no standard in the light of which one could reject the destruction of health in experiments performed on human beings (as those done by the Nazis in Ravensbrück and in many other concentration camps); there could also be no criterion to distinguish animal health from human health, or animal experiments from those performed on humans; both could be defended and neither could be rationally rejected. In fact, there would be no basis for supporting the ethical character of health care. Moreover, the state would have no foundation for its efforts to protect health, nor any limits to what individuals or groups could declare to be the value of health. Thus an economic disaster would result from the commitment to health care if there were no objective value-standards for its determination.

D. Is There an Objective Value-knowledge?

But even if one would accept an objectivist ontological position, further questions about the human capacity to actually grasp its content may arise. Hence, with Engelhardt one may ask whether the objective, positive content of health is actually open to our rational knowledge or accessible only through acts of human or religious beliefs. In other words, one could hold that the value of health is, ontologically speaking, objective but that we fail to possess any objective value-knowledge. Thus, answering the question about the inherent ontological objectivity or subjectivity of the value of health does not yet solve the epistemological question of whether there is knowledge of objective values available to us without the mediation of faith.

While all the authors of this volume agree on the existence of *ontological* objectivity, some disagree with any ability of human reason to detect an objective 'species-plan,' or an objective value thereof. This is Engelhardt's position, who combines epistemological relativism with ontological objectivism. It is chiefly in relation to this aspect of the contemporary debate on the nature of health that Donohue-White,

Cuddeback and Seifert seek to identify rational ways of recognizing objectivity. And they do so by various means:

- by pointing out aspects of health which presuppose objective standards and criteria, so that their denial will lead to self-contradictions in any theory of health;
- by reference to examples whose evidence is as such undeniable;
- by showing the immediate and direct evidence of general essential facts about human health that are grounded on necessary distinctions within the notion of value and good; and
- by *reductio ad absurdum,* showing that nobody can seriously deny objective standards in the determination of health and unhealth, without having to accept absurd consequences.

The conclusion these authors derive from these arguments is that the subject's capacity of knowing the objective value of health cannot be denied without falling into contradictions.

II. ANTHROPOLOGICAL ISSUES

Returning now to the WHO definition of health, it seems quite evident that it refers to the value of specifically *human* health, which alone can be defined in terms of "complete physical, mental and social well-being." (WHO, 1946). This volume can be read as an extensive interpretation of the personalist implications of the WHO definition of health, precisely with regard to this specifically human and humane understanding of health. Our book represents an attempt to bring out the specifically personal aspects of the concept of health which this definition certainly presupposes, but does not specify. The authors of this volume make an effort to state precisely what the human person's health is, in its distinction to the health of other living beings, such as plants or animals.

A. Overcoming Reductionism

A first aspect central to the WHO definition of health as well as to the contributions gathered in this volume is the need for an, anthropologically speaking, non-reductionist conception of human health that definitively overcomes previous mechanistic, materialistic, and exclusively biomedical conceptions of human health. This is certainly an aspect of crucial significance. For it is clear that immense consequences follow from whether or not we conceive of human health in purely mechanistic and physiological terms or also include under the notion of the well-being of persons the psychological, mental, spiritual and social dimensions of the person.

All the contributors of this volume agree on the fact that health in general, and human health in particular, cannot be reduced to the functioning of a machine, even though a correct functioning of the optical, mechanical, chemical, and electrical processes of the body do actually constitute part of human health, as Seifert argues. Indeed, this fact explains in part the success of previous mechanistic conceptions of health. But even among recent attempts to provide non-reductionist conceptions of human health – as those grounded on the General System Theory – we find some problematic aspects, as the critical investigations by Taboada prove.

There are profound differences with respect to the account of individuality and personhood given by different schools and authors. In fact, some theories that appear to be non-reductionist turn out in the end to reduce the datum of personal health and of the human person to a non-personal equilibrium of purely material factors. Such emergent theories of health cannot truly account for human health, as Ide and Taboada show. Already the irreducibility of life and soul to matter proves the irreducibility of health and well-being to purely material functions. Specifically human health, above and beyond its prebiological and biological aspects and conditions, includes also aspects that reach much farther than the biological health of plants or animals. Its proper concept requires an adequate understanding of the important dimensions of personal individuality and communality. These aspects of human health prove wholly irreducible to material functions, as discussed in Ide's, Roa's, Seifert's and Taboada's contributions.

Hence, by including the mental, spiritual and social dimensions of the person in the concept of health, the nature of the person's health can be conceived in totally different ways regarding its roots and causes, Ide and Roa argue. Indeed, if one conceives health as a pure consequence of electrical or biochemical events in the person's body, then the nature, causes, and cures of health problems need to be necessarily conceived in a reductionist way. Their conditions and causes do not yet guarantee a properly personalist understanding of health.

B. The Need for Personalist Categories

In particular, the five papers gathered in the first section of the book deal
with the question of whether human health requires new categories with
respect to those fit for describing health in general. In their affirmative
answer to this question, i.e., in their insisting on the need for a new and
properly human concept of health, a second significant consensus of the
contributions of this volume is manifest. The idea that some specifically
human and personalist categories need to be explicitly addressed in any
objective approach to the nature of human health and health care is
probably one of the most significant results of the philosophy of health
presented in this volume.

Regarding the philosophical question as to which pre-biological,
biological, and specifically personal dimensions account for human
health, we find much dispute among members of the medical profession
and among philosophers. Nevertheless, this question is of the utmost
importance, both theoretically (for a basic understanding of the human
person), and practically (for the way in which one should treat human
beings). For example, a purely biological concept of mental health would
lead to treating all psychic conflicts with drugs or chemical substances.
Instead, if with Roa and Seifert we accept that mental health possesses a
link to intentionality, structural rationality, freedom and the moral sphere,
an appropriate treatment of mental problems would require also an
acquisition of knowledge, of meaning, of an adequacy of our conscious
relation to objects, etc. This could be achieved, for instance, by
counselling. Thus, the methods of cure of mental health cannot be
reduced to mere pharmacological treatments.

Counselling becomes an important and specifically personalist method
of providing medical and psychological help to persons. It addresses a
person as a rational being capable of understanding and of making a free
contribution towards coping with critical situations and towards the
process of health-recovery. Humanist schools of psychology (particularly
logotherapy) and programs of drug rehabilitation which appeal to free
choices of patients presuppose such a non-reductionist concept of mental
health. Their conception of what constitutes an appropriate method for
curing human persons is built thereupon. It is interesting to realize that
this conception of health and cure was already present in ancient
philosophy of medicine, as Reale shows in his paper.

It is precisely by affirming the mental, spiritual and social dimensions of the person and her health that the authors of this volume provide an objective foundation for a non-reductionist and specifically personalist concept of human health.

C. The Nature of the Human Person as the Measure of Health

Another key issue in the current debate on the nature of human health is the question about the norm and normativity of health. Is the content of human health – because it is not a property of impersonal things but the health of free agents – open to any kind of interpretation, or does the health of a conscious subject possess an objective nature grounded in the objective 'species-plan' or essence of the human person? Does objectivity refer here only to statistics, or is it possible to go beyond the empirical method and detect some teleological structure and purposiveness of human nature which allows us to have some objective criteria to judge what is healthy and what is unhealthy in human beings?

Any theory of human health – tacitly or implicitly – refers to a standard or norm of health. Many authors today – particularly Boorse (1975, 1977) – argue that the standard of health depends on a 'species-plan' and is far from being whimsically determined by social, historical or political definitions. He believes, however, that the norm of such a species-plan is the statistical frequency of certain characteristics of living organisms. Nordenfelt (1987, 1993, 1993a), on the other hand, points out that there are many deviations from statistical values which are in no way unhealthy (for example, the extraordinary bodily strength and other features of sports champions that are frequently signs of a higher degree of health or of a functioning of the body above the average and not symptoms of unhealth).

Upon closer consideration, there are two aspects of the question about the source of such criteria for human health that are frequently confused and yet distinct: one is ontological, the other epistemological. Since we already referred to some of the epistemological aspects of this debate in the section above, we shall concentrate here on its ontological dimension. Ontologically speaking, an objective foundation of the criteria for human health rests on the assumption that the teleology, nature, values, order, etc. that are required for determining physical and mental health exist objectively and independently from arbitrary self-determination.

This ontological assumption leads to a rejection of any theory of human health based exclusively on the idea that health can be measured by the harmony between a subject's life-goals and the capabilities of achieving them. Even the addition of the qualification 'fundamental' or 'basic' to these goals – as Nordenfelt proposes – will not do, although it goes already in the direction of assuming some human nature as a standard for the healthiness of such goals themselves. For how can fundamental life-goals be distinguished from non-fundamental ones without any reference to that which makes them fundamental? To root their fundamental nature in nothing but the subject's desire to accomplish them will certainly not do and leads to a circular argument.

To avoid such a circular argument, we must look for some standard of health in human nature or in the essence of a living being, a standard that cannot be derived from mere statistics. Two of the contributions to this volume specifically address this aspect. Donohue-White and Cuddeback acknowledge that a comprehensive understanding of health includes both value and normative dimensions, and argue for a an objective normativity of health rooted in human nature. Lavados shows through examples that health judgments include a reference to an ideal norm, and raises the question of what kind of norm is referred to here and how it is determined. He rejects the idea of statistical normality as inadequate to the experience of disease and instead suggests the concept of 'clinical normality', i.e., the sense of the normal and abnormal which the physician presupposes in clinical practice. In clinical practice the physician establishes a correlation between an observed defect and the foreseen impact on the individual's overall functionality.

However, Lavados argues, this notion of clinical normality presupposes the philosophical notion of nature: "the normal is the natural," it refers to "a mode of functioning [which] conforms to the natural design of that kind of organism" (p. 200, in this volume). Being healthy corresponds to functioning in conformity with the natural design of the organism; disease consists in a state wherein the organism is unable to exercise one or more of the functions typically performed by a member of the species. The conclusion is that one needs to refer to a deeper measure or criterion for health and unhealth than statistics can provide.

The term 'nature' refers here to a discernible, general design or structure that determines an individual as a member of the human species. This design is understood as a natural kind and not merely as a statistical norm. Health is the state in which the specific actualization and flourishing of this nature can be achieved. Thus, the criterion according to which we recognize whether the development of a human person constitutes a flourishing or a decay, health or unhealth, is the person's own nature. In other words, it is the nature of the human person which fundamentally determines health and unhealth.

However, given the nature of the human person as a free agent, the person's essential plan is much more than a species-plan as we encounter it in other living beings. It involves intellectual, affective, and volitional dimensions that need to be developed. For this reason, we need to unfold here a 'personalistically' conceived essential plan. This plan coincides with the vocation of human beings to realize their personal nature and to become what they are called to be, as Ide shows. At the same time, at the level of the human person we have not only a general nature but also the individual character of each single person. This means that at the level of the person, we encounter a completely new type of 'species plan'. This plan is not simply carried out by nature and instincts as in spiders or birds. The human person's essence cannot be actualized without rational and free acts which also play a significant role for personal health. As Seifert and Roa have argued, to realize this sphere of personal creativity, a sphere largely absent from the health of animals, belongs also to a person's health. And this gives rise to art, language, symbols, etc., i.e., to the enormous riches of human culture.

D. On the Limits of the Concept of Health

This leads us organically to a further crucial question related to the standards for determining what constitutes the health of the human person, namely the problem of circumscribing its limits. Is health the state of "complete well-being" of the person as the WHO definition argues, or is this too utopian and too encompassing an understanding of health, which must be corrected by inserting health into an organic context of goods and goals of human persons?

Many of the contributions to this volume attack precisely such a utopian concept of health. Buttiglione, Pasquini, and van Spijk, for instance, introduce the distinction between basic health care and a more encompassing sense of health and health care. They also introduce the foundations for other delimitations between health and higher goods in the context of which health is situated and to which it is subordinated. Both from a theoretical and from a practical point of view the totalitarian and utopian moments in the WHO definition of health and the political and practical dangers ensuing from them are discussed by these authors.

This point is of the highest theoretical interest in arriving at an adequate grasp of human health but also of great practical interest. For in order to limit the promotion of health and the duty of persons with respect to it, society and state, as well as individuals, will have to have a clear concept of its limits. The authors of this volume seem to combine two significant results of their investigation in this regard: on the one hand, they point out the vastness of the content of health, including also its aesthetic, mental, spiritual and social dimensions. On the other hand, they struggle with arriving at a clear standard for distinguishing those goods of the human person which are higher than health itself. In so doing, they introduce sound criteria for the limits of health without which health-care would become a wholly unmanageable task, both morally and politically.

Of the many aspects of the contemporary debate on the concept of health that have been addressed in this book, it is perhaps this latter issue which, in spite of the convincing hints in Buttiglione and Pasquini's contribution, would require a further critical reflection that could lead to another book on 'health and its limits.' There are of course many other aspects of the question about the nature and value of human health treated in this volume that open fascinating paths for further research. We hope that our efforts may serve to raise the interest of others who will collaborate in this interesting task.

International Academy of Philosophy
Principality of Liechtenstein

REFERENCES

Boorse, C.: 1975, 'On the distinction between disease and illness,' *Philosophy and Public Affairs* 5.
—— 1977, 'Health as a theoretical concept,' *Philosophy of Science* XLVI.
Gadamer, H.G.: 1987, 'Apologie der Heilkunst,' *Gesammelte Werke Bd. 4: Neuere Philosophie II*, J.C. Mohr (Paul Siebeck), Tübingen.
Mackie, J. L.: 1977, *Ethics: Inventing Right and Wrong*, Penguin Books, New York.
Nordenfeldt, L.: 1987, *On the Nature of Health*, Reidel, Dordrecht.
—— 1993, 'Concepts of health and their consequences for health care,' *Theoretical Medicine* 14(4).
—— 1993a, *Quality of Life, Health and Happiness*, Avebury, Aldershot, Brookfield, Hong Kong, Singapore, Sydney.

World Health Organization (WHO): 1946, *Proceedings and Final Acts of the International Health Conference*, held in New York from 19 June to 22 July 1946. *Off. Rec. Wld Hlth Org.*, 2, 1-143.

NOTES ON CONTRIBUTORS

Rocco Buttiglione Ph.D. is Full Professor of Philosophy of Politics, Society and Economics of the International Academy of Philosophy in the Principality of Liechtenstein, and Member of the European Parliament in Brussels.

Patricia Donohue-White Ph.D., M.phil., lic. S. theol., is Assistant Professor of Theology and Philosophy at the Franciscan University of Steubenville, Steubenville, Ohio, and was a research assistant at the International Academy of Philosophy (IAP) in the Principality of Liechtenstein in the SNF-project that led to this volume.

H. Tristram Engelhardt, Jr. is Professor Emeritus in the Departments of Medicine, Community Medicine, and Obstetrics and Gynecology, at Baylor College of Medicine, and Professor in the Department of Philosophy at Rice University in Houston, Texas.

Kateryna Fedoryka Cuddeback holds a Ph.D. in Philosophy from the International Academy of Philosophy in the Principality of Liechtenstein and a B.A. in English Literature from the Christendom College, Front Royal, VA. She is a consultant and free-lance writer on population control and was a research assistant at the International Academy of Philosophy in the Principality of Liechtenstein in the SNF-project that led to this volume.

Pascal Ide holds a Ph.D. in Medicine and Philosophy and an M.A. in Theology. He is Official in the Vatican Congregation for Catholic Education.

Manuel Lavados, M.D., Ph.D., is Assistant Professor in the Center for Bioethics at the Medical Faculty of the Pontifica Universidad Católica de Chile, in Santiago, Chile.

Manuela Pasquini holds an M.A. in Philosophy from the International Academy of Philosophy in the Principality of Liechtenstein.

Giovanni Reale is Full Professor of Ancient Philosophy at the Catholic University of Milan, Director of the Centro di Ricerche di Metafisica (Sezione di Metafisica e Storia della Metafisica), and Director of the International Center for the Study of Plato and the Platonic Roots of Western Philosophy at the International Academy of Philosophy in the Principality of Liechtenstein.

Armando Roa, M.D., was Professor of Psychiatry and Director of the Center of Bioethics and Humanistic Studies at the Medical School of the Universidad de Chile, and President of the Instituto de Chile and of the Academia Chilena de Medicina.

Josef Seifert, Ph.D., is Rector and Full Professor of Philosophy – with special emphasis in Epistemology, Metaphysics, and Philosophical Anthropology – of the International Academy of Philosophy in the Principality of Liechtenstein, and was Director of the SNF research project that led to this volume.

Piet van Spijk, is a Medical Doctor (FMH Internal Medicine) in Luzern, Switzerland.

Paulina Taboada, M.D., M.phil., is Assistant Professor in the Department of Internal Medicine, and Executive Director of the Center for Bioethics at the Medical Faculty of the Pontificia Universidad Católica de Chile, in Santiago, Chile. She was a research assistant at the the International Academy of Philosophy in the Principality of Liechtenstein in the SNF-project that led to this volume.

INDEX

Philosophy and Medicine

21. G.J. Agich and C.E. Begley (eds.): *The Price of Health*. 1986
ISBN 90-277-2285-4
22. E.E. Shelp (ed.): *Sexuality and Medicine*. Vol. I: Conceptual Roots. 1987
ISBN 90-277-2290-0; Pb 90-277-2386-9
23. E.E. Shelp (ed.): *Sexuality and Medicine*. Vol. II: Ethical Viewpoints in Transition.
1987 ISBN 1-55608-013-1; Pb 1-55608-016-6
24. R.C. McMillan, H. Tristram Engelhardt, Jr., and S.F. Spicker (eds.): *Euthanasia
and the Newborn*. Conflicts Regarding Saving Lives. 1987
ISBN 90-277-2299-4; Pb 1-55608-039-5
25. S.F. Spicker, S.R. Ingman and I.R. Lawson (eds.): *Ethical Dimensions of Geriatric
Care*. Value Conflicts for the 21th Century. 1987 ISBN 1-55608-027-1
26. L. Nordenfelt: *On the Nature of Health*. An Action-Theoretic Approach. 2nd,
rev. ed. 1995 SBN 0-7923-3369-1; Pb 0-7923-3470-1
27. S.F. Spicker, W.B. Bondeson and H. Tristram Engelhardt, Jr. (eds.): *The Contra-
ceptive Ethos*. Reproductive Rights and Responsibilities. 1987
ISBN 1-55608-035-2
28. S.F. Spicker, I. Alon, A. de Vries and H. Tristram Engelhardt, Jr. (eds.): *The Use
of Human Beings in Research*. With Special Reference to Clinical Trials. 1988
ISBN 1-55608-043-3
29. N.M.P. King, L.R. Churchill and A.W. Cross (eds.): *The Physician as Captain of
the Ship*. A Critical Reappraisal. 1988 ISBN 1-55608-044-1
30. H.-M. Sass and R.U. Massey (eds.): *Health Care Systems*. Moral Conflicts in
European and American Public Policy. 1988 ISBN 1-55608-045-X
31. R.M. Zaner (ed.): *Death: Beyond Whole-Brain Criteria*. 1988
ISBN 1-55608-053-0
32. B.A. Brody (ed.): *Moral Theory and Moral Judgments in Medical Ethics*. 1988
ISBN 1-55608-060-3
33. L.M. Kopelman and J.C. Moskop (eds.): *Children and Health Care*. Moral and
Social Issues. 1989 ISBN 1-55608-078-6
34. E.D. Pellegrino, J.P. Langan and J. Collins Harvey (eds.): *Catholic Perspectives
on Medical Morals*. Foundational Issues. 1989 ISBN 1-55608-083-2
35. B.A. Brody (ed.): *Suicide and Euthanasia*. Historical and Contemporary Themes.
1989 ISBN 0-7923-0106-4
36. H.A.M.J. ten Have, G.K. Kimsma and S.F. Spicker (eds.): *The Growth of Medical
Knowledge*. 1990 ISBN 0-7923-0736-4
37. I. Löwy (ed.): *The Polish School of Philosophy of Medicine*. From Tytus
Chałubiński (1820–1889) to Ludwik Fleck (1896–1961). 1990
ISBN 0-7923-0958-8
38. T.J. Bole III and W.B. Bondeson: *Rights to Health Care*. 1991
ISBN 0-7923-1137-X

Philosophy and Medicine

Philosophy and Medicine

KLUWER ACADEMIC PUBLISHERS – DORDRECHT / BOSTON / LONDON